Innovation in Africa

Innovation in Africa

Levelling the Playing Field to Promote Technology Transfer

FERNANDO DOS SANTOS

OXFORD
UNIVERSITY PRESS

Great Clarendon Street, Oxford, OX2 6DP,
United Kingdom

Oxford University Press is a department of the University of Oxford.
It furthers the University's objective of excellence in research, scholarship,
and education by publishing worldwide. Oxford is a registered trade mark of
Oxford University Press in the UK and in certain other countries

© Fernando dos Santos 2024

The moral rights of the author have been asserted

First Edition published in 2024

All rights reserved. No part of this publication may be reproduced, stored in
a retrieval system, or transmitted, in any form or by any means, without the
prior permission in writing of Oxford University Press, or as expressly permitted
by law, by licence or under terms agreed with the appropriate reprographics
rights organization. Enquiries concerning reproduction outside the scope of the
above should be sent to the Rights Department, Oxford University Press, at the
address above

You must not circulate this work in any other form
and you must impose this same condition on any acquirer

Public sector information reproduced under Open Government Licence v3.0
(http://www.nationalarchives.gov.uk/doc/open-government-licence/open-government-licence.htm)

Published in the United States of America by Oxford University Press
198 Madison Avenue, New York, NY 10016, United States of America

British Library Cataloguing in Publication Data

Data available

Library of Congress Control Number: 2023948052

ISBN 978-0-19-285730-9

DOI: 10.1093/oso/9780192857309.001.0001

Printed and bound in the UK by
Clays Ltd, Elcograf S.p.A.

Links to third party websites are provided by Oxford in good faith and
for information only. Oxford disclaims any responsibility for the materials
contained in any third party website referenced in this work.

This book is dedicated to my father António Machado dos Santos (in memoriam) and my mother Maria Manuel Macuácua; my wife Nilza dos Santos; and my kids Nadia, Sheila, Kiara, and Alessio, for each one of them provided me with the unwavering support, guidance, care, and love needed in the different stages of my life.

Preface

It is a matter of fact that no country has ever developed without putting the necessary emphasis on technological progress and industrialization. In view of Africa's underdevelopment and technological deficit, the United Nations General Assembly proclaimed 20 November 'Africa Industrialization Day' through Resolution A/RES/44/237, in December 1989. Further, the General Assembly adopted, in 2016, Resolution A/RES/70/293, proclaiming 2016–2025 as the Third Industrial Development Decade for Africa (IDDA III). These political commitments demonstrate not only Africa's dire technological needs, but also that the international community is aware of this shortfall and that it is undertaking concrete, though non-binding, commitments to address technological deficit in Africa. Central to industrial development is undoubtedly access, use, and absorption of technology. Despite the fact that Africa pioneered technology development and made significant contributions to global science, the continent continues to lag behind as a producer of technology and has been relegated to a mere consumer of it: because of poverty, it is financially inhibited in accessing modern and effective technologies except where allowed by the rights-owners, as access to Covid-19 vaccines demonstrated.

This is where the fundamental topic of transfer of technology comes into play. In fact, countries which lack the ability to produce their own technology are compelled to obtain it through technology transfer mechanisms. Having observed the strong link between technology transfer and progress of nations, some attempts have been made globally in the past to establish an international regime to regulate flows of technology benefiting developing states, but without much success. The United Nations Conference on Trade and Development (UNCTAD) developed a Draft International Code of Conduct for the Transfer of Technology to remove barriers on the acquisition of technology by developing countries which are imposed by multinationals currently dominating the international technology market. The Draft Code was abandoned in 1986 due to a lack of consensus and since then fragmentation has ensued. Over eighty international instruments and numerous sub-regional and bilateral agreements that contain provisions dealing with technology transfer have been adopted worldwide. Further, some proposals to adopt an international legal instrument or a model contract on technology transfer were tabled at the World Intellectual Property Organization (WIPO) but did not yield results.

A compelling attempt to address the issue of promotion of technology transfer and fostering innovation for the benefit of Least Developed Countries (LDCs) was devised in the context of Articles 7 and 66.2 of the TRIPS Agreement. Article 66.2

of the TRIPS Agreement requires developed states to provide a set of incentives to promote and encourage technology transfer by companies and institutions located in their own jurisdictions to LDCs. It is argued that facilitating technology transfer for the benefit of LDCs forms part of the bargain that persuaded those states to agree to set up strong intellectual property (IP) systems during the negotiations of the TRIPS Agreement. However, this provision has not received much attention, hence its implementation has largely been ignored, denying LDCs the ability to establish the much-desired sound and viable technological base for their development.

This book is an attempt to fill the existing gaps in the institutional and scholarly analysis and bring to the fore the important role of technology transfer and innovation in assisting LDCs on the African continent to build their technological capabilities. The book will therefore provide practical tools for African countries to shape their position regarding the matter in the negotiation forums at WIPO and the World Trade Organization (WTO). Additionally, it will offer suggestions to African LDC governments to undertake some initiatives to attract foreign technologies and promote the development of local technologies in response to local needs.

In order to create a conducive environment for technology transfer and innovation in Africa, the book first proposes a reinterpretation of the philosophical justifications of the IP system based on considerations of justice, charity, and cosmopolitanism as perfected by the rights-based approach. Subsequently, it offers solutions to overcome the challenges currently posed by the inefficiencies in the flow of technologies to LDCs and the fragmentation of the international technology transfer regime. Notably, the book proposes the maximization of the implementation of the TRIPS Agreement's provisions related to technology transfer and the adoption of an international legal instrument which is designated: Agreement on Trade-Related Issues of Technology Transfer and Innovation (TRITTI), to be administered by the WTO. The book also advocates that LDCs be required to undertake proactive efforts to set up an enabling environment for technology transfer to thrive through the establishment of the right mechanisms and incentives, attracting technologies into their respective states, facilitating technological learning, and the fostering the development of absorptive capabilities, which will finally catapult African LDCs along the route of innovation.

Acknowledgements

The inspiration to write this book resulted from my research during my doctoral studies at the University of Witwatersrand in Johannesburg, South Africa. My supervisor, Professor Malebakeng Agnes Forere, strongly encouraged me to put my thoughts into writing, highlighting the importance of Africa-centred scholarship.

A special appreciation to Professor Irene Calboli from the Texas A&M University School of Law, a prolific scholar who also emphasized the need to see more studies springing up from the African continent and who guided me towards the right platform that might be interested in my work.

Professor Caroline Ncube from the University of Cape Town has been always very supportive, and her encouragement gave me more confidence to undertake and conclude this work.

I would also like to thank Doctor Marisella Ouma, a truly African respected intellectual property law scholar, for always cheering me and providing valuable guidance whenever it was needed.

A big thank you to Professor Lee Stone from the University of South Africa (UNISA) for her valuable inputs in the contents of this book.

My recognition also goes to the anonymous reviewers of the initial proposal for this book, who enabled the initiative to go through and provided important insights that shaped its structure and content.

A heartfelt appreciation to my wife Nilza dos Santos and to my kids, Nadia, Sheila, Kiara, and Alessio, for understanding the many days and hours of absence in order to undertake research and distil it into this book.

Contents

List of Boxes   xvii
List of Abbreviations   xix

PART I. FOUNDATIONS

1. General Introduction   3
   A. Africa's Challenges in Accessing and Exploiting Technology for Development   3
      1. The Importance of Technology for Development   3
      2. Technology Development in Africa and the Defining Role of its Protection   6
      3. Current Initiatives to Promote Technological Development in Africa   9
      4. The Challenges to Shaping a Legal Framework to Promote Technology Transfer at the Global Level   13
      5. Structure of the Book   17

2. The First Manifestations of the Ingenuity of Humankind and its Protection   19
   A. Introduction   19
   B. The First Manifestations of Humankind's Ingenuity and the Mechanisms for its Protection   21
   C. The World's Scientific and Technological Progress   22
      1. The Contribution of Africa   22
         a) Inputs into scientific and technological progress   22
         b) Africa's technological 'conservatism'   27
      2. The Contribution of Asia   30
      3. The Contribution of Europe   31
   D. The Birth of Intellectual Property   34
      1. Requirements for the Development of the Intellectual Property System   34
      2. Political and Legal Framework for the Protection and Promotion of Intellectual Property   35
         a) African context   35
            i) Political environment   35
            ii) Legal framework   37
            iii) Individual ownership of knowledge   39
         b) The Asian context   41
            i) Political environment   41
            ii) Legal framework   41
            iii) Individual ownership of knowledge   42

xii CONTENTS

   c) The European context   44
    i) Political environment   44
    ii) Legal framework   45
    iii) Individual ownership of knowledge   46
  E. Conclusion   48

3. Dynamics of the Extension of the Western-Style IP System into the African Continent   51
  A. Introduction   51
  B. Pre-colonial Extension of the Intellectual Property System to Africa   53
   1. Protection of Knowledge and Innovation in Pre-colonial Africa   53
   2. Rationale of the Introduction of Western-Style Intellectual Property Systems in Africa   54
   3. Modalities of Introduction of the Intellectual Property Systems in Africa   56
   4. The Administration Structures   59
  C. The Intellectual Property System in the 'Non-colonized' African States   60
   1. Context   60
   2. The Case of Liberia   61
   3. The Case of Ethiopia   63
  D. Post-Independence Implementation of Intellectual Property Systems in Africa   67
   1. The Role of Intellectual Property in Africa's Post-Independence   67
   2. Africa's Post-Independence Industrialization Policies and Intellectual Property   68
   3. Intellectual Property in the Context of the Current Industrialization and Innovation Policies   73
    a) Kenya   74
    b) Nigeria   75
   4. Some African Best Practices in Mainstreaming Intellectual Property Policies into National Development Programmes   77
    a) Policymaking space for IP policy development   77
    b) Rwanda's Intellectual Property Policy   78
    c) Mozambique's Intellectual Property Strategy   79
    d) South Africa's Intellectual Property Policy   80
    e) South Africa substantive examination priority areas   81
  E. Conclusion   82

PART II. THE USE OF INTELLECTUAL PROPERTY TO FOSTERING DEVELOPMENT IN AFRICA

4. The Challenges of Promoting Innovation through Technology Transfer into Africa   87
  A. Introduction   87

B. Philosophical Debates on the Reinterpretation of Locke's Labour
   Theory to Facilitate Transfer of Technology and Foster Innovation ... 89
C. Fragmentation in the Legal Framework on Technology Transfer ... 95
   1. The United Nations Covenant on Economic, Social and Cultural
      Rights Framework and Technology Transfer ... 95
   2. The UNCTAD Draft International Code of Conduct for
      the Transfer of Technology ... 96
   3. The United Nations Convention on the Law of the Sea and
      Transfer of Technology ... 98
   4. The United Nations Framework Convention on Climate Change
      Regime of Technology Transfer and the Kyoto Protocol ... 99
   5. The Convention on Biological Diversity and Transfer of Technology ... 100
   6. The Vienna Convention and the Montreal Protocol on
      Substances that Deplete the Ozone Layer and Technology Transfer ... 101
D. Proposed Reforms on the International Framework for
   Regulating Technology Transfer ... 102
   1. The Proposed Multilateral Agreement on Access to Basic
      Science and Technology under the WTO Framework ... 102
   2. The Proposed Implementation of ABST under the WIPO
      Development Agenda ... 104
   3. Proposal to Establish an International Technology Transfer
      Agreement under the WTO ... 107
E. Conclusion ... 109

5. Maximizing the Use of the TRIPS Agreement to Promote
   Technology Transfer and Innovation in Africa ... 111
   A. Introduction ... 111
   B. Clarifying Article 66.2 of the TRIPS Agreement on Technology
      Transfer ... 112
      1. Overview of Article 66.2 ... 112
      2. The Duty-Bearers ... 113
      3. The Nature of Incentives to be Granted ... 115
         a) Incentives to companies and institutions in developed states to
            promote transfer of technologies ... 115
         b) Establishing a mechanism to assess the impact of incentives on
            transfer of technology ... 119
         c) The proposal for the establishment of clusters of technology
            transfer ... 121
      4. The Use of Information and Communication Technologies ... 123
   C. Institutional Arrangements to Support the Implementation of
      Article 66.2 ... 124
      1. National Institutions to Monitor the Implementation of Article 66.2 ... 124
      2. Impact-Assessment Studies on the Implementation of Article 66.2 ... 126
      3. The Proposal for a WTO Advisory Centre for Technology Transfer
         and Innovation ... 132

xiv CONTENTS

    D. Enforcement of Article 66.2 through the WTO Dispute Settlement Mechanism   134
        1. The WTO Dispute Settlement Mechanism and TRIPS   134
        2. Challenges in Exploiting the Potential of the Dispute Settlement Mechanism to Enforce the Claims of LDCs   136
    E. Conclusion   138

PART III. LEVELLING THE PLAYING FIELD TO PROMOTE TECHNOLOGY TRANSFER AND INNOVATION IN AFRICA

6. The Proposal to Establish the Agreement on Trade-Related Issues of Technology Transfer and Innovation (TRITTI)   143
    A. Introduction   143
    B. Rationale for the Establishment of the Treaty   145
    C. Addressing Challenges Raised by the Failed Attempts at Adopting an International Framework   147
        1. UNCTAD Draft Code Stumbling Blocks   147
        2. Scope   148
        3. Applicable Law   150
        4. Legal Character of TRITTI   152
        5. Dispute Settlement   153
        6. Technical Assistance   155
        7. WTO as a Host Organization   156
        8. Measures to Promote Collaboration of the Private Sector in the Transfer of Technologies   158
    D. Conclusion   160

7. The Role of Developing States and LDCs in the Quest for Technology Transfer: Recommendations   163
    A. Introduction   163
    B. The Development of Technology Absorption and Adaptation Capabilities as Enablers of Technology Transfer in LDCs   165
        1. Development of Absorptive Capacities in Africa   165
        2. Government Interventions to Develop Absorptive Capacity   166
            a) Government institutional frameworks focused on science, technology, and innovation   167
            b) Improvement of funding for research and development   171
            c) Development of human capital to enable technological progress   173
            d) Adequate exploitation of science, technology, engineering, and mathematics (STEM)   178
            e) Socialization or contextualization of science and technology   181

   f) Technological development through acquisition and
    indigenization of foreign technologies   182
     i) Utility models as a tool for indigenization of foreign
      technologies   182
     ii) The acquisition and indigenization of foreign technologies
      through utility models in East Asian states   186
     iii) Indigenization of foreign technologies through the use of
      utility models in Africa   188
   g) Leveraging the digital economy to fast-track Africa's
    innovative capabilities   190
   h) Focus on leveraging the Fourth Industrial Revolution   197
 C. Conclusion   202

8. Final Conclusions and Recommendations   205
 A. Introduction   205
 B. Summary of the Findings and Recommendations   207
  1. Overcoming the Philosophical Obstacles to Africa's
   Participation in the Global Innovation Process and
   Access to Technology   207
  2. Maximizing the Use of Article 66.2 to Promote Technology
   Transfer to Africa   211
  3. The Proposal for Establishing an International Treaty on
   Technology Transfer   213
  4. Enabling Environment to Facilitate Flows of Technology in
   Recipient States   214
 C. Conclusion   218

References   221
Index   241

List of Boxes

Box 5.1 Guidelines for the Field Impact-Assessment of the Technologies
Transferred in the Context of the Implementation of Article 66.2 of TRIPS   128

Box 7.1 National Strategy to Promote the Use of Utility Models   191

List of Abbreviations

4IR	Fourth Industrial Revolution
ABST	Agreement on Access to Basic Science and Technology
ACTTI	Advisory Centre for Technology Transfer and Innovation
ACWL	Advisory Centre on WTO Law
AFCTA	African Continental Free Trade Area
AIDA	Accelerated Industrial Development of Africa
ARIPO	African Regional Intellectual Property Organization
ASTT	Advisory Service on Transfer and Development of Technology
AU	African Union
CBD	Convention on Biological Diversity
CDIP	Committee on Development and Intellectual Property
COMESA	Common Market for East and Southern Africa
CTCN	Climate Change Technology Centre and Network
DSM	Dispute Settlement Mechanism
EAC	East African Community
ECCAS	Economic Community of Central African States
ECOWAS	Economic Community of West African States
EPO	European Patent Organization
EU	European Union
FDI	Foreign Direct Investment
ICESCR	International Covenant on Economic, Social and Cultural Rights
ICJ	International Court of Justice
IMF	International Monetary Fund
IP	Intellectual Property
IPRS	Intellectual Property Rights
ISI	Import Substitution Industrialization
LDCs	Least Developed Countries
NDVI	Normalized Difference Vegetation Index
NEPAD	New Partnership for Africa's Development
NIEO	New International Economic Order
OAPI	African Intellectual Property Organization
OAU	Organisation of African Unity
OECD	Organisation for Economic Co-operation and Development
PAIPO	Pan African Intellectual Property Organization
PIS	Patent Information System
RECs	Regional Economic Communities
R&D	Research and Development
SADC	Southern African Development Community
STEM	Science, Technology, Engineering and Mathematics

LIST OF ABBREVIATIONS

STI	Science, Technology and Innovation
STISA-2024	African Union Science, Technology and Innovation Strategy for Africa 2014–2024
TFTA	Tripartite Free Trade Area
TISC	Technology and Innovation Support Center
TRIPS	Agreement on Trade-related Aspects of Intellectual Property Rights
TRITTI	Agreement on Trade-Related Issues of Technology Transfer and Innovation
UAVs	Unmanned Aerial Vehicles
UN	United Nations
UNCTAD	United Nations Conference on Trade and Development
UNECA	United Nations Economic Commission for Africa
UNFCCC	United Nations Framework Convention on Climate Change
UNIDO	United Nations Industrial Development Organization
WIPO	World Intellectual Property Organization
WTO	World Trade Organization

PART I
FOUNDATIONS

1
General Introduction

A. Africa's Challenges in Accessing and Exploiting Technology for Development

1. The Importance of Technology for Development

The value of technology in promoting development and improving productive capacities and people's lives has been long and widely recognized.[1] Technological progress has contributed immensely to improving productivity, reducing costs, lessening the burden of hard work on humans, and spreading more benefits to them. Indeed, it is also undisputable that the goal of technological development is to provide higher standards of living for humans by making available to them products and services that meet their demands and needs.[2] Arguably, it is only when any given society has achieved a certain level of satisfaction of human needs that is considered developed. This view is embedded in the etymological origin of the concept of technology. Indeed, this term derives from the ancient Greek 'technologia', a composite word constituted by the term 'techné' meaning 'art, skill' and the suffix 'logia' meaning 'words, speech'. It follows that technology is a means through which humans satisfy their needs and desires by studying 'tech-niques' and using inventions and discoveries for the overall good of human development. Viewed in this perspective, technology plays a central role in the socio-economic development of the society.

This strong belief in a pervasive and defining role for technology in modern society is known as technological determinism.[3] Although twentieth-century scholars have coined the term, its seeds were planted by the enthusiasm deriving

---

[1] Engwa Azeh Godwill, 'Science and Technology in Africa: The Key Elements and Measures for Sustainable Development' (2014) 14 GJSFR: G BIO-TECH & GEN 16, 21; Keith Maskus, 'Encouraging International Technology Transfer' [2004] IPRSD 1, 7; Anja Breitwieser and Neil Foster, 'Intellectual Property Rights, Innovation and Technology Transfer: A Survey' (The Vienna Institute for International Economic Studies 2012) 88, 47 <https://wiiw.ac.at/p-2646.html> accessed 24 April 2023; Surendra J Patel, 'The Technological Dependence of Developing Countries' (1974) 12 J ModAfrStud 1, 4.
[2] Satyajit Rath, 'Science and Technology: A Perspective for the Poor' (1994) 29 EPW 2916, 2917; Godwill (n 1) 18.
[3] Merritt Roe Smith and Leo Marx, *Does Technology Drive History?* (The MIT Press 1994); Jill Lepore, 'Our Own Devices: Does Technology Drive History?' *The New Yorker* (5 May 2008) <www.newyorker.com/magazine/2008/05/12/our-own-devices> accessed 28 June 2018.

Innovation in Africa. Fernando dos Santos, Oxford University Press. © Fernando dos Santos 2024.
DOI: 10.1093/oso/9780192857309.003.0001

from the Industrial Revolution in Europe which spread its tentacles in America.[4] Theorists backing this view derive from inventions such as the compass, the printing press, and the computer a clear demonstration of the transformative and impactful power of technology to society. Advocates of this approach, reflecting the viewpoint of American society, conceive technology as a panacea for all challenges that humanity faces and believe that it is only from this that the well-being of humanity can be assured.[5] In general, those that rally behind technological determinism believe that there is agency in technology that can drive history and social development.

Western countries have long embraced this view and have taken advantage of this important tool to advance their own societies, rendering them much more developed than other nations. For example, it has been established that the sustainable and long-term economic growth of many developed countries was generated by technological changes that enabled development of new products, processes, or industries.[6] Notably, the introduction of technology and its agency in infrastructure, energy, industry, health, education, communication, finance, entertainment, transport, agriculture, and environment has transformed the daily life of people and shaped society in a way never before seen and at an impressive pace.

This trend was also witnessed recently in the East Asian nations, where betting on the use of technology enabled them to be co-opted into the process of production spurred by globalization.[7] Invariably, the East Asian countries that achieved a considerable level of progress are those that relied upon the adoption of modern technology in their economic fabric to steer the course of their developmental processes.[8] The examples of countries such as China, Hong Kong, Indonesia, Korea, Malaysia Singapore, Thailand, and Taiwan bear testimony to the importance of the adoption of technology not only to fast-track development but also to develop competitiveness in order to actively participate in the global market.

In general, economic and social development has been linked to technological development.[9] Similarly, from the 1930s to the 1980s, the concept of innovation in developed countries has been largely viewed as synonymous with technological innovation.[10] Indeed, notwithstanding some trade-offs and

---

[4] Smith and Marx (n 3).
[5] ibid.
[6] Linsu Kim, 'Technology Policies and Strategies for Developing Countries: Lessons from the Korean Experience' (1998) 10 TechnoAnalStrateg 311, 311.
[7] Larry E Westphal, 'Technology Strategies for Economic Development in a Fast Changing Global Economy' (2002) 11 EconInnovNewTechnol 275, 275–320.
[8] ibid.
[9] Maskus (n 1) 7; Breitwieser and Foster (n 1) 47.
[10] Directorate-General for Research and Innovation (European Commission) and others, *Social Innovation as a Trigger for Transformations: The Role of Research* (Publications Office of the European Union 2017) 11 <https://data.europa.eu/doi/10.2777/68949> accessed 24 April 2023; Peter RA Oeij and others, 'Understanding Social Innovation as an Innovation Process: Applying the Innovation Journey Model' (2019) 101 J BusRes 243, 243–54.

scepticism,[11] technological progress generally appears to be well placed to provide a significant contribution to the improvement of the living standards of people in developing countries and lift them out of poverty.[12]

The crucial role of technology in promoting the development of nations has been established and is therefore undeniable. However, there is a lack of balance in the availability and share of technology throughout the world, with rich countries of the North experiencing a surplus while developing countries are struggling to access it.[13] In the early 1970s, Patel claimed that there was indeed a storehouse of technological expertise, accumulated in the developed countries since the Industrial Revolution, but observed that most of mankind located in other parts of the world continued to remain ill-fed, ill-clad, ill-housed, and illiterate.[14] Patel also claimed that if the available technological knowledge were transferred to developing countries, their socio-economic transformation could happen rapidly. It appears evident that, there is spoilage of knowledge and technology, which is abundant in the developed world that is not reaching the developing world. For instance, technology that is necessary to meet the basic needs of the developing world, such as electricity production, water and sanitation, agriculture, communication, and several areas of basic health, is widely available in developed states, but the gap in terms of access and use of that technology is widening between developed states and Least Developed Countries.[15]

---

[11] The belief that mere transfer of technology will enable developing countries to leap across the centuries and repeat the Industrial Revolution is refuted by Norman Clark and Calestous Juma, 'Technological Catch-Up: Opportunities and Challenges for Developing Countries' 3 <www.academia.edu/58329996/Technological_Catch_Up_Opportunities_and_Challenges_for_Developing_Countries> accessed 24 April 2023.

[12] To counter the negative effects of the exaggerated focus on technological innovation, the concept of 'social innovation', born in the eighteenth century, is gaining ground as a corrective measure, as highlighted by the European Union (n 10) 17. Indeed, it has been observed that technological progress alone has some trade-offs, such as fuelling unemployment, easing gender-based discrimination, worsening working conditions, and raising health and safety issues and negative environmental impacts: Isabella Massa, 'Technological Change in Developing Countries: Trade-Offs between Economic, Social, and Environmental Sustainability' [2015] MERIT WP 7-22; Mariacristina Piva, 'The Economic Impact of Technology Transfer in Developing Countries' (2004) 112 RISS 433, 433–69 also explains that: 'negative consequences might lead to adverse competitive effects with domestic firms, displacement of workers, and negative welfare implications'.

[13] Nsongurua J Udombana, 'The Third World and the Right to Development: Agenda for the Next Millennium' (2000) 22 HRQ 753, 783.

[14] Patel (n 1) 1–5.

[15] UNCTAD, *Transfer of Technology and Knowledge-Sharing for Development: Science, Technology and Innovation Issues for Developing Countries* (United Nations 2014) 3; Udombana (n 13) 13.

2. Technology Development in Africa and the Defining Role of its Protection

Similarly to what happened in other continents, manifestations of innovation and creativity occurred in Africa at the early stages of development of humankind.

Nonetheless, technology has been marginalized in the processes of production, which has led Africa to consistently rely on the human factor in its agricultural, transport, and manufacturing activities, with sporadic use of rudimentary technological implements.[16] This trend was influenced by a cultural construct of risk aversion. The 'mythology of risk' froze opportunities for the sharing of technical information, blocked imitation, and stifled innovation. Therefore, although the old African kingdoms and empires recognized and promoted the work of creators and innovators, they failed to develop mechanisms to foster innovation in a consistent way.[17] Risk aversion seems to be embedded in the African culture, associating technological innovations with high risk, thus justifying hesitancy in embracing them.[18] Until the present day, technology conservatism creates fear of the unknown and delays adoption and adaptation of innovations in the African context.[19] Low literacy is a further contributing factor in slowing the development or penetration of technology. Indeed, there is a proven strong correlation between literacy and critical-innovative modes of thought.[20] Low rates of literacy prevailing on the continent have had a severe impact on the ability to adopt new or foreign technologies to date.

These findings provide enough evidence to assert that social components can play a pivotal role in the development or adoption of technologies. Cultural behaviour is certainly a determinant in the selection of technologies that are acceptable in the local context in view of their compatibility with local perceptions, customs, practices, and beliefs. Therefore, it can be safely stated that cultural inertia was decisive in setting a low pace in the adoption and improvement of technology and consequently blocking the technological progress of the African continent.

---

[16] Ralph A Austen and Daniel Headrick, 'The Role of Technology in the African Past' (1983) 26 ASR 163, 169.

[17] Mark C Suchman, 'Invention and Ritual: Notes on the Interrelation of Magic and Intellectual Property in Preliterate Societies' (1989) 89 ColumLawRev 1264, 1279; Paul Kuruk, 'Protecting Folklore Under Modern Intellectual Property Regimes: A Reappraisal of the Tensions Between Individual and Communal Rights in Africa and the United States' (1999) 48 AmUniLawRev 786 <https://digitalcommons.wcl.american.edu/aulr/vol48/iss4/2> accessed 28 October 2020.

[18] Austen and Headrick (n 16) 173.

[19] Similar pattern is also found in Asia: cultural factors and neo-Confucian philosophy led to the stagnation of the remarkable technological achievements of Chinese society, see John Alan Lehman, 'Intellectual Property Rights and Chinese Tradition Section: Philosophical Foundations' (2006) 69 JBE 1, 6–8; Ying Lowrey and William J Baumol, 'Rapid Invention, Slow Industrialization, and the Absent Innovative Entrepreneur in Medieval China' (2013) 157 PAPS 1, 15.

[20] Austen and Headrick (n 16) 16.

While the prevailing knowledge-governing systems in the pre-colonial era were unable to develop a legal framework to promote and protect technological progress in pre-colonial era in Africa and to promote use of technology, the colonial system did not try to improve the situation. On the contrary, the colonial system suppressed any chance of the local innovation system thriving by promoting western-style innovation protected by an IP system that was biased against indigenous knowledge. This colonial approach hindered the process of innovation and technological progress on the continent.

Reflecting this design, manufacturing activities were discouraged by the authorities during the colonial era.[21] Initial stages of manufacturing that happened in Africa around 1920 focused exclusively on the processing of agricultural raw materials for export, such as palm oil from Cameroon, Nigeria, and Ghana; lime juice from Ghana; sawmilling from Ivory Coast; and sisal, cotton, coffee, and tea from Tanzania.[22] Only very limited industrialization, having as its purpose simply to serve European elites living in the colonies, was introduced, such as baking, food processing, beverages, tobacco, textiles, shoes, and wooden products. Both international agents and local colonial governments viewed industrialization policies in the colonies as a threat to the established monopoly of production of manufactured goods which was reserved for the metropolis and the supply of cheap raw materials to the same market from the colonies.[23] Of relevance is the fact that industrialization was allowed within the colonial framework and the industries were all owned and managed by Europeans and to some extent by Asian and Middle Eastern nationals, while Africans were barred by law from owning companies, or else they lacked the necessary education, skills, and capital in several countries.[24] As a result, qualified human resources were scarce and limited to some professions and the public service.

In view of these circumstances, at the time of independence the prevailing scenario in African states was characterized by a primary exporting economy focused on minerals and crops, lacking an entrepreneurial class, with no capital, relying on unqualified people with incipient knowledge of technology.[25]

The dawn of decolonization sparked hope for change and the establishment of a new era that would finally take into account the interests of Africa. However, in reality, quite the opposite occurred. Indeed, the developmental policies adopted by African states after independence strived to meet the needs of their people by promoting rapid industrialization of their territories. This was because the prevailing

---

[21] William Steel and Jonathan Evans, *Industrialization in Sub-Saharan Africa—Strategies and Performance* (World Bank 1985)7.
[22] ibid.
[23] Ana Paula F Mendes, Mário A Bertella, and Rudolph FAP Teixeira, 'Industrialization in Sub-Saharan Africa and Import Substitution Policy' (2014) 34 BrazilJPolEcon 120, 121.
[24] Steel and Evans (n 21) 8.
[25] Mendes, Bertella, and Teixeira (n 23) 125.

scenario by the time that African states attained their independence in the 1950s and 1960s was characterized by a huge dependence upon, and strong link with, European metropoles with regard to export markets, technology, management and entrepreneurship capabilities, capital, and the economic system in general.[26] The economic system was therefore detached from the local reality and was designed to serve the interests and needs of the colonial masters instead of addressing the needs of local populations. Newly independent states reacted vociferously to this scenario and resolved to change their economies and explore new production activities with a view to tackle issues of relevance to their people. To that end, they followed the successful model adopted by Latin American states of import substitution as the strategy to overcome their challenges.[27] In carrying out import substitution, African states started with non-durable consumer goods, such as foodstuffs, drinks, tobacco, and textiles; intermediate goods, such as rubber, chemicals, oil, metal, and non-metallic mineral products; and capital goods, such as electrical machines and transportation equipment.[28] Ultimately, these states intended to shift the export structure from primary goods to value-added goods with higher prices.[29]

African governments assumed total control of the industrial development system and policies from the second half of the 1960s, with the intention of ensuring the implementation of their vision.[30] Governments undertook nationalization programmes and adopted import substitution industrialization to achieve their goals. These state-led development programmes rolled out in Africa in the 1960s and 1970s produced significant results, as evidenced by the growth in gross domestic product (GDP) and comprehensive social welfare programmes.[31] Seemingly, progress was happening, and this suggested that these policies were successful, given the exponential growth in the number of state-owned companies that provided clear evidence of the industrialization of countries such as Cameroon, Ghana, Ivory Coast, Kenya, Mali, Mauritania, Nigeria, Senegal, Tanzania, Uganda, Zambia, and Zimbabwe.

However, this positive picture soon faded as the growth in size was not followed by better production. Indeed, only Nigeria, Kenya, and Zambia truly became industrialized and, of these three, only Nigeria and Kenya maintained good performance.[32]

---

[26] Steel and Evans (n 21) 13.
[27] ibid 126.
[28] ibid 128; Overseas Development Institute, *Industrialization in Sub-Saharan Africa* (1986).
[29] Mendes, Bertella, and Teixeira (n 23) 126.
[30] ibid 129–30.
[31] 'Capitalism—Africa—Independence, State-Led Development, and Import-Substitution Industrialization—African, Economic, Women, and Program—JRank Articles' <https://science.jrank.org/pages/8525/Capitalism-Africa-Independence-State-Led-Development-Import-Substitution-Industrialization.html> accessed 9 August 2017.
[32] Mendes, Bertella, and Teixeira (n 23) 131.

Among many obstacles to the successful implementation of these policies were lack of human capital and little knowledge of technology.[33] For example, the agricultural sector did not benefit from public investment in technological innovation, and farmers continued growing crops using outdated practices and technology, rendering them unable to support industrial growth.[34] Apparently, governments did not focus on issues related to technological innovation; consequently, although the introduction of new technologies to improve and increase production and distribution of goods and services was found to be relevant, no specific technological policy and planning were set out to guide the initiatives undertaken.[35] The result of that lack of vision led to high technological dependence on foreign know-how, to the extent that the domestic factor endowments of the country were grossly neglected.[36]

In assessing this context, it appears evident that the desire to achieve rapid industrialization prompted African states to opt for procurement of technology in developed states and to establish capital-intensive industry. As a result, technology importation became an end in itself, and no effort was made to enable the countries to absorb the technology. Therefore, although modest results of industrialization were initially achieved, ownership of the process and transfer of skills and technology did not occur. This is due to the fact that governments did not focus on creating local capacity to sustain the industrialization process but were merely concerned with the production of non-durable consumer goods that could substitute imports and satisfy local and ephemeral needs. Such a model could not be sustainable in the long run and collapsed within decades.

3. Current Initiatives to Promote Technological Development in Africa

The value of technology and innovation in promoting development and improving productive capacities and people's lives cannot be overemphasized.[37] Accordingly, policymakers, scientists, the private sector, and civil society have recognized that science, technology, and innovation are the drivers of African development in the knowledge era. Many ventures have been embarked on to promote technological development in Africa, both at the national and continental levels.

---

[33] ibid.
[34] ibid.
[35] ibid 133.
[36] Louis N Chete and others, 'Industrial Policy in Nigeria: Opportunities and Challenges in a Resource-Rich Country' in Carol Newman and others (eds), *Manufacturing Transformation: Comparative Studies of Industrial Development in Africa and Emerging Asia* (Oxford University Press 2016); Udo Ekpo, 'Nigeria Industrial Policies and Industrial Sector Performance: Analytical Exploration' (2014) 3 IOSR JEF 1, 2–4.
[37] Godwill (n 1) 21; Joe McMahon and Margaret A Young, 'The WTO'S Use of Relevant Rules of International Law: An Analysis of the Biotech Case' (2007) 56 ICLQ 907, 910; Maskus (n 1) 7; Breitwieser and Foster (n 1) 47; Patel (n 1) 4.

The discussion of technological development initiatives at the national level is limited in this instance to Kenya and Nigeria because of their apparent success in developing their industrial sectors in recent years. Indeed, efforts were made in recent years to pair innovation and industrial policies. For example, Kenya adopted the National Industrialization Policy Framework for Kenya, 2012–2030 in November 2012.[38] The policy aspires to transform Kenya into a rapidly industrializing middle-income state and focuses on value addition for both primary and high-value goods. It also aims to develop linkages between industrial sub-sectors and other productive sectors to drive the industrialization process and provide strategic direction for the growth and development of the respective sectors.[39] The policy defines three industrial sectors as priority areas for development: labour-intensive, medium to high-technology sectors, and advanced manufacturing technologies.[40] Kenya has also adopted innovation policies, with the first policy being launched in 2006. In 2009, a comprehensive policy on Science, Technology, Innovation Policy and Strategy (STIPS) was developed for purposes of mainstreaming the application of science, technology, and innovation in all sectors and processes of the economy.[41] Furthermore, Kenya has enacted the Science, Technology and Innovation Act No 28 of 2013, the objectives of which are threefold: to facilitate the promotion, coordination, and regulation of the progress of science, technology, and innovation in the country; to assign priority to the development of science, technology, and innovation; and to entrench science, technology, and innovation into the national production system and for connected purposes.[42] The Act sets out important and transformative measures such as: imposing an obligation on public universities and research institutions to enact internal IP policies, establishing transfer of technology or innovation offices, requiring the government to commit to funding research activities by at least 2 per cent of the GDP, and setting up institutions for the promotion of innovation, namely the National Commission for Science and Technology, the National Research Fund, the National Innovation Agency, and the National Museum for Science and Technology.

---

[38] Ministry of Industrialization, 'National Industrialization Policy Framework for Kenya 2012–2030: Transforming Kenya into a Globally Competitive Regional Industrial Hub' <https://repository.kippra.or.ke/handle/123456789/1037> accessed 18 June 2017.

[39] ibid. Policy objective 4.7 on the standards, quality infrastructure, and intellectual property rights.

[40] The three priority areas include: labour intensive sectors such as agro-processing, textiles and clothing, leather and leather goods; medium- to high-technology sectors, namely iron and steel, machine tools and spares, agro-machinery and farm implements, and pharmaceuticals; and, as a third area, advanced manufacturing technologies, which includes biotechnology and nanotechnology, ibid 25–32.

[41] Bitange Ndemo, 'Effective Innovation Policies for Development: The Case of Kenya' in *The Global Innovation Index 2015: Effective Innovation Policies for Development* (WIPO 2015).

[42] Science, Technology and Innovation Act 2013.

Moving on to Nigeria, the World Bank data indicates that the country hosts the largest economy in Africa.[43] In 2023, its population was more than 223 million,[44] with a GDP of over US$500 billion. However, half of the GDP is generated by primary sectors such as agriculture, while the oil and gas sector is the major driver of the economy, to the extent that, in 2011, it made up over 95 per cent of export earnings and about 85 per cent of government revenue.[45] To change this scenario the government adopted an agenda for economic transformation known as 'Nigeria Vision 20:2020', a long-term plan for stimulating Nigeria's economic growth and launching the country onto a path of sustained and rapid socio-economic development. The industrialization strategy incorporated in this vision aimed to achieve greater global competitiveness in the production of manufactured goods by linking industrial activity with primary sector activity, domestic and foreign trade, and service activity. The vision also incorporated elements of relevance to science, technology, and innovation. 'Nigeria Vision 20:2020' is currently being revised to ensure continuity and efficiency in the country's development planning and to reflect the country's vision of development for 2030.[46] The Federal Ministry of Finance, Budget, and National Planning is also leading an ongoing initiative to develop the '2050 Long-Term Vision for Nigeria' (LTV-2050) aimed at making Nigeria a socio-economically advanced nation with a technologically enabled, digitally connected, diversified, and inclusive sustainable economy.[47] Previous attempts to provide a direction in Nigeria's economic and technological transformation may also be spotted in the National Science and Technology Policy formulated in 1986 with the aim of placing science and technology at the centre of national development. The policy emphasized the importance of promoting transfer of technology from foreign to local firms.[48] The Nigeria Industrial Revolution Plan (NIRP), adopted in 2014, is also pertinent; it was labelled as the first strategic, comprehensive, and integrated roadmap to industrialization, designed as a five-year plan to accelerate the development of industrial capacity in Nigeria. The plan aimed to increase manufacturing's contribution to GDP from 4 per cent to 6 per cent by 2015, and ambitiously, above 10 per cent by 2017.[49] Available data reveals that although there

---

[43] Charles Yao Kouame and others, 'WDI—Many African Economies Are Larger than Previously Estimated' (2019) <https://datatopics.worldbank.org/world-development-indicators/stories/many-economies-in-ssa-larger-than-previously-thought.html> accessed 18 June 2021.

[44] 'World Population Dashboard—Nigeria' (United Nations Population Fund) <www.unfpa.org/data/world-population/NG> accessed 17 June 2017.

[45] Chete and others (n 36) 36.

[46] Terhemba Daka, Geoff Iyatse, and Femi Adekoya, 'Nigeria Abandons Vision 20:2020, Dreams Agenda 2050' The Guardian Nigeria News—Nigeria and World News (10 September 2020) <https://guardian.ng/news/nigeria-abandons-vision-202020-dreams-agenda-2050/> accessed 25 April 2023.

[47] '2050 Long-Term Vision for Nigeria (LTV-2050)' <https://unfccc.int/documents/386681> accessed 27 April 2023.

[48] Chete and others (n 36) 123.

[49] 'Nigerian Industrial Revolution Plan (NIRP)' <www.nipc.gov.ng/product/nigerian-industrial-revolution-plan-nirp/> accessed 10 November 2018.

was substantial improvement, oil still dominates Nigerian GDP and the contribution of manufacturing remained below 10 per cent in 2018.[50] The Nigeria Industrial Revolution Plan is based on the desire to drive a process of intense industrialization, founded on sectors where Nigeria has comparative advantage.[51]

At the African Union level, several documents have been adopted with the aim of spurring the technological development of the continent, namely Agenda 2063, the AU Science, Technology and Innovation Strategy for Africa (STISA-2024), and the Accelerated Industrial Development of Africa (AIDA).

The current overarching document related to development is the Agenda 2063,[52] which recognizes science, technology, and innovation (STI) as multi-functional tools and an enabler for achieving continental development goals. The document contains objectives, some timeframes, enablers, and a call to action for the realization of the aspirations of African people.

Another important guiding document developed at the continental level is STISA-2024.[53] This document places STI at the epicentre of Africa's socio-economic development and growth. The Foreword of the Strategy highlights that, with the advent of STISA-2024, the African Union possesses a wonderful tool to accelerate the continent's transition to an innovation-led, knowledge-based economy.[54] STISA-2024 aims at driving STI to make an impact on critical sectors defined by the AU, such as agriculture, energy, environment, health, infrastructure development, mining, security, and water. The priorities identified in this strategy are as follows: the eradication of hunger and the achievement of food security, prevention and control of disease, communication (physical and intellectual mobility), protection of our space, living together and building society, and wealth creation.[55] Specifically, on wealth creation, STISA-2024 recognizes that to accelerate Africa's transition to an innovation-led, knowledge-based economy, the focus must be on empowering human resources with the necessary skills and greater emphasis must be placed on innovation and on the appropriate adaptation of technology and existing research results. The Strategy also advocates for the promotion of innovative technologies for processing African natural resources.[56]

---

[50] 'Nigeria's Economy: Services Drive GDP, but Oil Still Dominates Exports' (*Africa Check*, 2018) <http://africacheck.org/fact-checks/reports/nigerias-economy-services-drive-gdp-oil-still-dominates-exports> accessed 12 March 2020.

[51] Such as the agro allied sectors; metals and solid minerals related sectors; oil and gas related industries; as well as construction, light manufacturing, and services.

[52] African Union, 'Agenda 2063: The Africa We Want' <https://au.int/en/agenda2063/overview> accessed 4 June 2022.

[53] African Union, 'Science, Technology and Innovation Strategy for Africa (STISA-2024)' <https://au.int/en/documents/20200625/science-technology-and-innovation-strategy-africa-2024> accessed 7 October 2017.

[54] ibid 9.

[55] ibid 22–23.

[56] ibid.

The AIDA Plan considers industrialization as the critical engine of economic growth and development and the essence of development in itself. AIDA was adopted in 2008 by the heads of state and government and contains seven Programme Clusters[57] to tackle priorities that need to be addressed to promote the coherent industrial development of Africa.[58] The Plan recognizes that the industrialization of Africa can only be accelerated and sustained on a solid technological base.[59]

4. The Challenges to Shaping a Legal Framework to Promote Technology Transfer at the Global Level

Developing countries, including those in Africa, acknowledge that technology is the main tool required to kindle its development, and hence need to explore ways to improve its uptake.[60] However, this technology is not freely available for developing countries, including several nations on the African continent. Notwithstanding all the legal and policy framework that was crafted to foster technological progress, this is still a mirage for the continent and the crude reality is that Africa is trailing behind other regions of the globe in the uptake of technology.

This situation has long been recognized and many initiatives have been undertaken by several organizations to address it. These initiatives premised on the fact that nations which lack the capacity to develop their own technology are only left with the option of borrowing it, mainly from private corporations, often from the developed world through technology transfer mechanisms.[61] Technology transfer is thus of paramount importance for the realization of the benefits of technological

---

[57] The seven programme clusters are: Industrial Policy and Institutional Direction; Upgrading Production and Trade Capacities; Promote Infrastructure and Energy for Industrial Development; Human Resource Development for Industry; Industrial Innovation Systems, R&D and Technology Development; Financing and Resource Mobilization; and Sustainable Development, see African Union, 'AIDA—Accelerated Industrial Development for Africa' <https://au.int/en/ti/aida/about> accessed 17 January 2017.

[58] These include: Policy on Product and Export Diversification, Natural Resources Management and Development; Infrastructure Development; Human Capital Development and Sustainability, Innovation, Science and Technology; Development of Standards and Compliance; Development of Legal, Institutional and Regulatory Framework; and Resource Mobilization for Industrial Development, African Union ibid 2.

[59] ibid 4.

[60] Breitwieser and Foster (n 1) 47; Hans Duller, 'Role of Technology in the Emergence of Newly Industrializing Countries' (1992) 9 ASEAN EcoBul 45, 45–54; Saon Ray, 'Technology Transfer and Technology Policy in a Developing Country' (2012) 46 JDA 371; Maskus (n 1) 7; M Scott Taylor, 'TRIPS, Trade, and Technology Transfer' (1993) 26 CJE/Revue canadienne d'économie 625, 625–37; Godwill (n 1).

[61] Ray (n 60) 371; Bruce A Larson and Margot Anderson, 'Technology Transfer, Licensing Contracts, and Incentives for Further Innovation' (1994) 76 AJAE 547, 547; Shane Tomlinson and others, *Innovation and Technology Transfer Framework for a Global Climate Deal An E3G Report with Contributions from Chatham House* (Chatham House 2008).

progress and plays a major role in facilitating the necessary inflows of technology to developing countries in general, and African countries in particular.

The important role played by technology to promote development has led to some attempts to establish an international regime to regulate technology transfer. One such initiative is the United Nations Conference on Trade and Development (UNCTAD) Draft International Code of Conduct for the Transfer of Technology.[62] UNCTAD developed a Draft International Code of Conduct to remove constraints imposed, due to domination of the international technology market by multinationals, on the acquisition of technology by developing states. The Draft Code proposed liberalization of trade in technology and the introduction of guidelines on the terms and conditions of transfer of technology to developing states.[63]

Notwithstanding protracted negotiations that lasted from 1976 until 1985, the Draft Code was never adopted by the General Assembly due to a lack of consensus.[64] Since the UNCTAD Draft Code was abandoned in 1986, new areas of public interest that were not considered in the negotiations have emerged, such as health, agriculture (protection of new varieties of plants), and climate change.[65] For these new areas, international legal instruments provide for their own separate regimes of technology transfer. Consequently, there are currently around thirty multilateral agreements that deal with technology transfer, half of which are related to global environmental protection.[66] UNCTAD identified over eighty international instruments and numerous subregional and bilateral agreements that contain provisions related to technology transfer.[67] Some of the relevant international instruments that contain provisions on technology transfer include the United Nations Convention on the Law of the Sea of 1982, the Montreal Protocol on Substances that Deplete the Ozone Layer of 1987, the Convention on Biological Diversity (CBD) of 1992, and the United Nations Framework Convention on Climate Change (UNFCCC) of 1992.

The proliferation of sectoral regimes in the technology transfer arena seems to denote a situation of fragmentation, as evidenced by the lack of a unitary or

---

[62] UNCTAD Secretariat, 'Draft International Code of Conduct on the Transfer of Technology' <https://digitallibrary.un.org/record/86199> accessed 13 May 2018.
[63] UNCTAD, *Transfer of Technology* (United Nations 2001).
[64] Susan K Sell, *Power and Ideas: North-South Politics of Intellectual Property and Antitrust* (State University of New York Press 1997).
[65] Padmashree Sampath and Pedro Roffe, 'Unpacking the International Technology Transfer Debate: Fifty Years and Beyond' (2012) <https://papers.ssrn.com/abstract=2268529> accessed 20 June 2021.
[66] Zhong Fa Ma, 'The Effectiveness of Kyoto Protocol and the Legal Institution for International Technology Transfer' (2012) 37 JTT 75, 77.
[67] UNCTAD, *Compendium of International Arrangements on Transfer of Technology: Relevant Provisions in Selected International Arrangements Pertaining to Transfer of Technology* (United Nations 2001).

harmonized law.[68] Such a scenario is detrimental to the countries which need to access technology because they have no legal instrument to back their request.

In a nutshell, the African continent finds itself in a dilemma, due to lack of the internal capacity to develop or to absorb technology necessary for its own development and the absence of legal means to claim support to access the much-needed technology. Notwithstanding all efforts to establish a coherent international regime to regulate technology transfer, the lacuna persists, hence the proposals put forward in this book. A two-pronged approach is suggested for the improvement of the existing international legal instruments that govern transfer of technology and innovation, or for the establishment of a new unified legally binding international instrument. This is to be coupled with other initiatives to boost local innovation undertaken by governments in developing countries, especially in Africa.

This book provides more—and implementable—input into the debate relating to the empowerment of the African continent to play a more meaningful and beneficial role in the global innovation system. Accordingly, the book explores prevailing theories and the current legal regime regulating IP to support, both philosophically and legally, the quest for innovation for the African continent.

The central undertaking of the book involves an analysis of the extent to which the current IP legal framework prioritizes the establishment and consolidation of a strong IP protection system, while simultaneously perpetuating the exclusion of innovation that springs from Africa. Underlying this critical analysis is the determination of whether the existing framework can be transformed so as to advance Africa's quest for development and innovation. The issue is whether, if correctly implemented or improved—especially taking into account ideals of justice as put forward by Locke and the development theory—the current international legal framework could create an enabling environment for innovation in Africa. Building on the current theories that have previously been used to justify strong IP regimes, the book demonstrates how the same theories can be applied to justify limitation of the IP appropriations in favour of a more balanced regime that can foster technological progress based on universal principles of justice, fairness, and cosmopolitanism.

Cosmopolitanism has constituted the overarching methodology for this book, aiming at the articulation of a more equitable distribution of IP rights that have as their purpose to promote Africa's development and foster innovation.[69] This

---

[68] Christopher Wong and others, 'From Reality to Law: Sustainable Technology Transfer—An Outlook' in Hans Henrik Lidgard, Jeffery Atik, and Tú Thanh Nguyễn (eds), *Sustainable Technology Transfer: A Guide to Global Aid and Trade Development* (Kluwer Law International 2012) 294.
[69] Tamara Hervey and others, *Research Methodologies in EU and International Law* (Hart Publishing 2011) 47.

reinterpretation sought, in the first place, to inspire an ethical responsibility on the part of governments, citizens, and companies of developed states to provide access to technology to developing states and support their innovation endeavours based on the ideals of charity and cosmopolitanism.

The relevance of the book lies in its provision of concrete proposals to overcome the current stagnation of the debate concerning international regulation of transfer of technology and promotion of innovation as witnessed today. The book proposes solid philosophical and legal tools to empower the African continent to reclaim its space in the global innovation system. These proposals consist of measures to leverage the existing legal and institutional framework, more specifically the Agreement on Trade-related Aspects of Intellectual Property Rights (TRIPS Agreement); the establishment of a new global legal framework to promote transfer of technology and foster innovation; institutional arrangements to facilitate related processes; and mechanisms to simplify the settlement of disputes. Further, the book provides new ways of developing synergies in the implementation of these mechanisms. The proposals tabled ultimately reinforce the quest of the African states to participate in the knowledge-based economy and ongoing innovation processes through an exponentially fairer and more equitable IP system.

However, the book also provides a nuanced and honest view of the reasons behind the misfortunes of African states (which cannot be attributed solely to external factors) regarding the adoption, development, and use of technology. Certainly, cultural approaches and misguided policies have contributed to the slow pace of development and adoption of technologies and in many cases have stifled innovation in the continent. Therefore, this book also constitutes a self-introspection of the continent of Africa on the possible initiatives to be undertaken in order to create the appropriate enabling environment for technologies to flow into the respective states and, most importantly, to facilitate the development of technological capabilities that will accelerate technological learning and absorption of the acquired technologies.

Overall, the book has philosophical, policy, legal, institutional, and practical significance. A new approach to the philosophical justifications of IP enables their reinterpretation to back Africa's plea for innovation; at the policy level, the book proposes some tools to stimulate the development of appropriate policies that can promote transfer of technology and foster innovation within the national innovation ecosystems; within the legal framework, the book proposes concrete legal instruments, especially in the global arena, in order to level the playing field and facilitate flows of technology to the countries in need; at the institutional level, specific institutional arrangements are proposed both at the national and international levels to coordinate actions; and, at the practical level, various recommendations are unveiled that once implemented may provide a great contribution to improve technology transfer and innovation dynamics for the benefit of developing countries, especially in Africa.

5. Structure of the Book

This book is structured in three parts and comprises eight chapters in total. Part I introduces and provides an overview of the birth of ingenuity, ancient mechanisms of protection of knowledge, and the introduction of the IP system—with limited success—in the African continent. In particular, Chapter 1 introduces the relevant issues of this study. As indicated in this first chapter, some factors contributing to technological lag in Africa are internal, and Chapter 2 therefore discusses the first manifestations of ingenuity of humankind and its protection. It demonstrates that, although those first creative signs occurred in Africa, and the continent contributed immensely to the development of science, it then reached a stalemate. The factors that led to the stagnation of innovation while other countries advanced will be highlighted. Among those factors, it emphasizes how the inability to nurture and use IP—admittedly not due entirely to fault on its part—had a negative impact on technological progress and innovation on the continent. Chapter 3 delves into the analysis of the dynamics of the extension of the IP system in Africa. It unveils the haphazard and inappropriate modalities of introduction of the system in the African continent during the period of colonialism and its negative impact that derailed the technological progress there. The chapter illustrates the attempts to industrialize the newly independent states in Africa, which also failed to mainstream IP into the industrial and innovation policies. The chapter finally assesses the efforts to redress the situation through the development of IP policies encouraging the emulation of the limited success stories identified.

Part II endeavours to illustrate the efforts to develop and implement international rules to promote transfer of technology with a view to fostering innovation. Accordingly, Chapter 4 highlights, first, the challenges to promoting innovation through technology transfer and then focuses on the efforts of the international community to establish a legal regime that enables technology transfer for the benefit of disadvantaged states, including those in Africa. Fragmentation of the international legal regime of transfer of technology and innovation is analysed and attention is drawn to its negative impact. Subsequently, ways of addressing that challenge are explored, including through attempts to develop an omnibus treaty on technology transfer.

Chapter 5 reviews the deficiencies of the current IP system and explores how to level the playing field to promote technology transfer and innovation for the benefit of developing states (especially in Africa), focusing on the TRIPS Agreement. To that end, the chapter attempts to elaborate on how to maximize the use of Article 66.2 of the TRIPS Agreement. Indeed, it is argued that some substantial results are achievable if the implementation of this provision is improved. The chapter further proposes the review and reinterpretation of some of the TRIPS concepts, innovative institutional frameworks are as well as full exploitation of the Dispute Settlement Mechanism of the WTO.

Part III provides a two-pronged approach to overcome the challenges of transfer of technology and fostering innovation. In Chapter 6, the proposal to establish a uniform international legal instrument on the transfer of technology to promote flows of technologies to Least Developed Countries is unveiled. Thereafter, Chapter 7 explores the role that candidate recipient states of technology transfer can play in attracting technologies into their respective territories. Hence, the chapter discusses possible strategies, policies, and interventions that these states can adopt to develop technological capabilities that will create the necessary enabling environment to facilitate technology absorption and home-grown innovation. To conclude, Chapter 8 summarizes the main findings and focuses on interconnections between them with a view to upholding the principal thesis of this book. The chapter also provides recommendations for the improvement of the IP system so as to promote technology transfer and foster innovation.

The book is not a panacea for all issues related to access to technology, in particular by the African countries, but it certainly provides some insights that can contribute to change the prevailing scenario and sustain, to a certain extent, the legitimate claims of the continent to be a player in the global innovation ecosystem.

2
The First Manifestations of the Ingenuity of Humankind and its Protection

A. Introduction

During every discussion at international level, there is a recurring sidenote to help African countries implement the provisions of any treaty at stake owing to Africa's low levels of technological development. Central to this chapter is an endeavour to locate the place of African countries in the evolution of technological development, the question being, has Africa always been wanting in technology? How did Africans survive before the arrival of Europeans? This chapter aims to provide the evidence found in different parts of Africa to fill that gap. The factual background to these questions is that the development of stone artefacts in ancient times was the first expression of the human capacity to develop technology. This was a simple technology comprising stone tools for cutting, scraping, and carving and for other heavy-duty work known as lithic technology. There is enough evidence that attests to the fact that the lithic technology was first developed in Africa: the oldest known deposit of cut stones was found in Omo and the earliest lithic technology appears in records located at Kada-Gona that are dated 2.6 million years ago; the same technology was also identified in Kada-Hadar and Omo-Shungura dating to around 2.4–2.3 million years ago—these sites are all located in Ethiopia, in Africa. Unequivocal evidence still on the ground that attests to how Africans mastered the use of stone prior to the arrival of Europeans is provided by the presence of two impressive construction sites: the Pyramids of Egypt built in approximately 2500 BCE and the Great Zimbabwe ruins built between 1100 and 1450 CE.

Stone tools were subsequently replaced by iron and other metal tools such as bronze during the Iron Age as illustrated in this chapter. There is evidence that iron was particularly plentiful in many parts of West Africa: some communities, such as those in the village of Nok located in the current northern region of Nigeria, mined and worked iron at the time of Christ;[1] there was also iron mining in north-western Central African Republic dating to 2200–1965 BCE;[2] iron-using societies

---

[1] Daniel Chu and Elliott P Skinner, *A Glorious Age in Africa: The Story of 3 Great African Empires* (Repr edn, Africa World Press 1998) 19.
[2] Étienne Zangato and Augustin FC Holl, 'On the Iron Front: New Evidence from North-Central Africa' (2010) 8 JAfrArchaeol 7, 7–23.

Innovation in Africa. Fernando dos Santos, Oxford University Press. © Fernando dos Santos 2024.
DOI: 10.1093/oso/9780192857309.003.0002

in Eastern Africa, such as Engaruka, south of the present Kenya-Tanzania border were also identified.[3] Subsequent research has demonstrated that Early Iron Age industrial sites were widespread in Rwanda and Uganda, substantiating the fact that technology was very well mastered by many communities in a vast area of the continent.[4] Knowledge of iron working brought great changes and rapid development to the people of West Africa: it prompted more productivity in farming and brought great military advantages, allowing the conquest of people who were not yet using iron weapons. Empires such as Mali and Ghana founded their strength on the use of iron, and their cities and middle-sized villages were booming with craftsmen, woodcarvers, silversmiths, goldsmiths, coppersmiths, blacksmiths, weavers, tanners, and dyers, providing enough evidence of the fact that metal technology was being used.[5]

Based on this short background, which indicates that Africans were active in technological development, the question that is of interest to this book is: what caused African countries to stall in technological development while their peers in other continents such as Europe drove their technological advancements?

To unravel these issues, trace the origins of technology, follow its evolution and spread throughout the world, and further explain why Africa failed to adopt it while other geographical areas were successful in that endeavour, a historical methodology is used.[6] Comparative methods are then employed to analyse the socio-economic and political environments that were favourable to technological progress in some latitudes such as Europe or that slowed that pace in others such as Asia with a view to drawing possible explanations for the African phenomenon.[7]

Therefore, this chapter starts by tracing the location of the first manifestations of humankind's ingenuity and its spread in other geographical areas of the world; subsequently, it analyses the genesis of the world's scientific and technological progress, including the contribution of Africa, Asia, and Europe. Owing to the relevance of intellectual property (IP) as the main driver of innovation and creativity, the requisites for its birth and development are analysed in the three continents. Conclusions are thus derived on the reasons why it was possible for IP to thrive in Europe and correspondingly for technology to find a fertile soil there while failing to do so in Africa and Asia.

---

[3] Walter Rodney and Angela Davis, *How Europe Underdeveloped Africa* (Verso 2018) 63.

[4] Debra Shore, 'Steel-Making in Ancient Africa' in Ivan Van Sertima (ed), *Blacks in Science: Ancient and Modern* (Transaction Publishers 1991) 162.

[5] Chu and Skinner (n 1).

[6] Berg and Lune eloquently state on this matter: 'Historical research attempts to systematically recapture the complex nuances, the people, meanings, events, and even ideas of the past that have influenced and shaped the present', Bruce L Berg and Howard Lune, *Qualitative Research Methods for the Social Sciences* (8th edn, Pearson 2011) 159.

[7] Francesca Bray, 'Technics and Civilization in Late Imperial China: An Essay in the Cultural History of Technology' (1998) 13 Osiris 11, 13; Pamela O Long, 'Invention, Authorship, "Intellectual Property," and the Origin of Patents: Notes toward a Conceptual History' (1991) 32 TechCult 846, 848.

B. The First Manifestations of Humankind's Ingenuity and the Mechanisms for its Protection

The main expression of human culture is to change the surrounding environment to fit one's needs in order to survive. To achieve this, humans have adapted natural materials into useful tools such as stone artefacts. Adaptation by humans to nature allowed them to survive in hostile environments.[8] In particular, the making and using of tools by human beings was a clear sign of their intellectual potential and ability to develop, which distanced them from other primates.[9] Coppens asserts in this respect that:

> Man and tools became a couple, no tools without Man, no Man without tools.[10]

It therefore goes without saying that the human being as we know them today is also the result of deliberate adaptation to the harsh environment through the development and use of tools.

The development of stone artefacts in ancient times was the first expression of the human capacity to develop technology. Leakey affirms that:

> the African continent was the scene of the most significant steps that were ever taken in the whole animal kingdom—the step of beginning to 'make tools to a set and regular pattern'.[11]

This is not surprising because if Africa is credited with being the original home of human species,[12] it is as a result home to the first attempts to develop some sort of technology in order to dominate nature. Indeed, if Africa is the original home of the human species, it is probable that the innovation and creativity that characterize human beings have also found their origin in Africa. In that regard, the development of stone artefacts on the African continent is evidence enough that

---

[8] Imre Bard, *History of the World: Africa and the Origins of Humans* (Steck-Vaughn 1992) 11–12.
[9] Rodney and Davis (n 3) 4; Jane Van Lawick Goodall, *My Friends: The Wild Chimpanzees* (1st edn, National Geographic Society 1967); Nancy Lawson, 'Man Not Only Toolmaker' (1964) 85 Science News <www.sciencenews.org/archive/man-not-only-toolmaker> accessed 18 June 2018.
[10] Yves Coppens, 'Hominid Evolution and the Emergence of the Genus Homo' (2013) 9 ScriVaria 121.
[11] Louis SB Leakey, *Progress and Evolution of Man in Africa* (Oxford 1961) 3; Eudald Carbonell and others, 'Eurasian Gates: The Earliest Human Dispersals' (2008) 64 JAR 195, 200; Coppens (n 10) 15. These sources suggest that the oldest known deposit of cut stones was found in Omo and Kada-Gona in Ethiopia, dating from 2.6 million years ago. This first generation of lithic technology was also found in two other Ethiopian sites, namely Kada-Hadar and Omo-Shungura. The sites of Lokalelei in Kenya and Senga in Congo consolidate the scenario with dates of around 2.4–2.3 million years ago.
[12] Claudio Tuniz, Giorgio Manzi, and David Caramelli, *The Science of Human Origins* (Routledge 2016) 14; Thomas O'Toole, 'The Historical Context' in April A Gordon and Donald L Gordon (eds), *Understanding Contemporary Africa* (3rd edn, Lynne Rienner Publishers 2001) 24; Sally McBrearty, 'The Origin of Modern Humans' (1990) 25 Man 129, 129–43; Coppens (n 10) 15.

unequivocally demonstrates that the first human attempt to develop technology, although rudimentary, was made in the continent of Africa.

The stone tools were then replaced by iron tools during the Iron Age.[13] Iron was particularly plentiful in many parts of Africa and there is evidence of iron-using societies in West and Eastern Africa.[14] Manufacture of metals was not limited to iron. Discovery of Ife bronze heads and Nok figurines in Northern Nigeria demonstrates that manufacturing of plastic art forms in West Africa precede the arrival of Europeans. It also shows that sophisticated working of some metals was in place well before any contact with western traditions.[15]

The use of metals, particularly iron, revolutionized the African continent and influenced the development of agriculture, as it accelerated effective forest clearance and garden cultivation, leading to more intensive food production and more productivity in farming.[16] There is also evidence of the development of a textile industry partly as a result of iron technology. The textile industry was therefore advanced in some regions of Africa, especially West and Central Africa, and produced unique pieces with elaborate designs to meet dressing requirements and styles unique to its local populations.[17]

The tool-making which continued throughout this time paved the way for the development of scientific and technological progress as described in Section C, 'The World's Scientific and Technological Progress'.

C. The World's Scientific and Technological Progress

1. The Contribution of Africa

a) Inputs into scientific and technological progress

The history of the evolution of technology in the world is well documented and described elsewhere and it will not be the focus of this book. Here, the fundamental

---

[13] The sequence Stone Age, Bronze Age, and Iron Age was first recognized at Copenhagen by CJ Thomsen between 1816 and 1865 and fits into the European context. It is disputed whether such a sequence can be applied in the African context as it seems that iron, bronze, copper, and gold coexisted side by side. In view of that, Goodwin suggests using the term 'Metal Age' rather than focusing on a single metal. He then defines the 'Metal Age' as 'that period manifesting a knowledge of the working of metal by methods which are or have become native to Africa, continuing through the introduction and successful assimilation of exotic elements and techniques, but ceasing with the historic period', see AJH Goodwin, 'Metal Age or Iron Age?' (1952) 7 SAAB 80, 80–82.

[14] Zangato and Holl (n 2) 7; Shore (n 4) 157.

[15] Chu and Skinner (n 1) 19.

[16] Roland Anthony Oliver and Brian M Fagan, *Africa in the Iron Age: C.500 BC–1400 AD* (Cambridge University Press 1975) 11, 18; Chu and Skinner (n 1) 19; Stanley B Alpern, 'Did They or Didn't They Invent It? Iron in Sub-Saharan Africa' (2005) 32 HistAfr 41, 74.

[17] John Thornton, 'Early Kongo-Portuguese Relations: A New Interpretation' (1981) 8 HistAfr 183; John Thornton and Andrea Mosterman, 'A Re-Interpretation of the Kongo-Portuguese War of 1622 According to New Documentary Evidence' (2010) 51 JAH 235.

task is instead to identify the building blocks of modern society and modern technology and to assess the extent of the contribution of Africa to the advancement of technological progress and to compare it with other regions of the globe, namely Europe and Asia. It is also relevant to analyse why Europe surpassed Africa and Asia as the latter two have declined in terms of their developmental dynamics.

It is worth introducing this topic by refering to the two most impressive construction sites erected in Africa before the arrival of the Europeans in the fifteenth century: the Pyramids of Egypt built in approximately 2500 BCE and the Great Zimbabwe ruins (as they are now known) built between 1100 and 1450 CE. What is remarkable about these sites is that their value is only conceivable if a combination of architecture and engineering skills is considered. It was the knowledge accumulated in those areas that paved the way for the development of the building of such great structures. Indeed, it reveals that, underlying this, there was already a high degree of development of science, technology, and managerial skills that could support the knowledge and logistics needed to develop such massive works.

By way of illustration, the great pyramid of Giza dates back to the thirtieth century BCE. This imposing and intriguing construction suggests that, by that time, Egypt could afford to gather together skilled craftsmen, mathematicians, physicians, architects, engineers, and managers to meticulously plan and execute such a huge structure.[18] Similarly, at least 800 years ago, south of the Sahara Desert, it was also possible to develop the massive stone complex of the Great Zimbabwe ruins.[19] The complex included nine separate stone sites and it is estimated that around 10,000 people were living in these enclosures, making it one of the largest cities of its time.[20] Building such complexes required sophisticated skills and industry. Great Zimbabwe was not an isolated structure: in a vast area which stretches between the current territories of Zimbabwe and Mozambique, more than 500 similar structures, although on a smaller scale, can be found.[21] The existence of similar structures in such a vast area attests once again to the knowledge that was common in those societies among the indigenous people in prehistoric days.

As we have highlighted, the achievements described were only possible because they were supported by solid scientific knowledge and, in particular, by a good understanding of physics and mathematics. Nevertheless, the contribution of Africa to the development of mathematics is yet to be recognized by western historians. However, great mathematicians such as Euclid, Claudius Ptolemy, Heron, Diophantus, and Theon were believed to be from Alexandria and therefore most probably Egyptian and it is because of them that the world inherited great works,

---

[18] Beatrice Lumpkin, 'The Pyramids: Ancient Showcase of African Technology' (1980) 2 JAC 10, 67.
[19] Molefi Asante and Kariamu Asante, 'Great Zimbabwe: An Ancient African City-State' (1983) 5 JAC 84, 84.
[20] ibid 85.
[21] Stanlake John Thompson Samkange and Tommie Marie Samkange, *Hunhuism or Ubuntuism: A Zimbabwe Indigenous Political Philosophy* (Graham Pub 1980) 16.

such as 'The Elements' (Euclid), 'The Almagest' (Claudius Ptolemy), 'Metrica' (Heron), and 'Arithmetica' (Diophantus).[22] A number of famous European mathematicians acquired their knowledge through interaction with the Muslims and Moors of North Africa.[23] As evidence, it is narrated that Adelard of Bath (1116–42 CE) made a long trip to Arab countries; and Fibonacci (Leonardo Pisano) (1170–240 CE) was initiated in mathematics while residing in a North African coastal city where his father was a merchant.[24]

Knowledge and use of mathematics in ancient Africa are not limited to Egypt. There are impressive traces of the use of mathematics in Africa, south of Egypt, such as the Yoruba system of numbers.[25] The Yoruba system expresses, for example, forty-five as 'five from ten from three twenties' therefore involving addition, subtraction, and multiplication just to express one number.[26] The system is extremely complex, denoting abstract reasoning rather than a simple observation of nature and interpretation.

Many other achievements in this area are attributed to African people, namely: technology to produce carbon steel attributed to Tanzanians; astronomical observation to the Dogons of Mali; proto-maths in South Africa and Eswatini; mathematical papyrus invented by the Congolese people; in addition to achievements in many other areas, such as language, mathematical systems, architecture, agriculture, cattle-rearing, navigation of inland waterways and open seas, medicine, and communication. Furthermore, Sona geometry (symmetrical and non-linear) which has become the basis of new mathematical ideas that mirror curves and various classes of matrices such as cycles, cylinders, and the helix were found in Angola; geometric patterns used in textiles, paintings, sculpture, cosmologies, architecture, and town planning found among the West and Central African people are the basis of the worldwide web; and methods of mathematical calculation that resemble the binary logic of mathematical calculation, similar to that which is behind today's Internet Explorer and other computer-based systems, were also found in Ethiopia and the Yoruba of Nigeria.[27]

In physics and physical science, the Greek philosopher Thales of Miletos is cited as the first scientist. That underestimates the fact that the first manifestations of physical science are embedded in the origins of human society and its efforts to solve their day-to-day challenges such as the development of throwing devices for spears; the invention of the bow and arrow; making flint weapons, tools, dwellings, boats; etc.[28] As extensively demonstrated throughout the chapter, all these

[22] Lumpkin (n 18) 105.
[23] ibid 107.
[24] ibid 105.
[25] Claudia Zaslavsky, 'The Yoruba Number System' in Ivan Van Sertima (ed), *Blacks in Science: Ancient and Modern* (Transaction Publishers 1991) 119.
[26] ibid.
[27] Mammo Muchie, 'Why Pan-African Education Should Be Promoted' (2015) 553 NewAfr 24–25.
[28] John Pappademos, 'An Outline of Africa's Role in the History of Physics' in Ivan Van Sertima (ed), *Blacks in Science: Ancient and Modern* (Transaction Publishers 1980) 180.

phenomena occurred first in Africa. Knowledge of measurement, mathematics, conversion of units of area, length, volume, weight, time, etc. were invented and developed as a logical method of ensuring survival and sustaining the development of society. The need for such knowledge and development of those scientific branches occurred naturally in the evolving societies in Africa. For the purposes of the present book, it should be understood that in order for that knowledge to promote technological progress and solve the daily challenges, it needs to be systematized and cannot be merely intuitive. Therefore, it may be concluded that the birth of the first principles of physical science occurred before Thales of Miletos, who up until now has been credited as being the first scientist, based on the prevailing western narrative.

Global science owes its origins to Egyptian concepts of the most fundamental physical quantities: distance, area, volume, weight, standards, units, and methods for accurate measurement of all of these quantities. Scientific measurement of time began as far back as 4241 BCE, when the Egyptians developed the calendar which is still in use today (with minor changes). The Egyptian calendar contained 365 days, twelve months of thirty days each, plus five festival days. The Egyptians also introduced the practice of starting the day at midnight and the names of the days of the week.[29] Therefore, credit must be given to Africans for having contributed immensely—indeed immeasurably—to the development of physical science. In view of that, Pappademos' assertion that the ability to conceptualize, to think abstractly, to generalize, and to discover new laws of physics was undoubtedly present in African people seems entirely plausible.[30]

The interest of Africans in science is also found in astronomy. Researchers Lynch and Robbins discovered a megalithic site dated 300 BCE in Namoratunga (meaning 'stone people') northwestern Kenya, with an alignment of nineteen basalt pillars that are non-randomly oriented towards certain stars and constellations, among them Aldebaran, Orion, Saiph, Sirius, Triangulum, Pleiades, and Bellatrix.[31] The pillars were aligned with the stars with extreme precision and minimal margin of error. This meticulous alignmemt and precision is once again a clear demonstration of the complexity of prehistoric cultural developments in Sub-Saharan Africa. The modern Cushites in East Africa have a calendar based on the rising of certain stars and constellations. Lynch and Robbins suggest that the accurate and complex calendar system of the Cushites may have been developed based on the knowledge of astronomy, already by the first millennium BCE in East Africa, of which Namoratunga is clear evidence.[32]

---

[29] ibid 187.
[30] ibid 194.
[31] B Mark Lynch and Lawrence H Robbins, 'Namoratunga: The First Archeoastronomical Evidence in Sub-Saharan Africa' in Ivan Van Sertima (ed), *Blacks in Science: Ancient and Modern* (Transaction Publishers 1980) 55.
[32] ibid.

One of the most amazing astronomical discoveries made by Africans and which continues to astound the scientific community is the in-depth grasp of the Sirius star-system by the Dogon of Mali. They knew that the Sirius star had an eliptical orbit around it. Although it took fifty years to complete, the Dogon even drew a diagram showing the trajectory of the star up until the year 1990 and showed that it was a heavy star although compacted.[33] Knowledge of the Sirius star-system by the Dogon is more than 700 years old and it appears that this knowledge was also shared by other black people in West Africa. The Dogon used to organize a ceremony in honour of Sirius every sixty years, when the orbits of Jupiter and Saturn converged. A wooden mask called the *Kanaga* dating from the thirteenth century, which was used by the Dogon to celebrate Sirius-related Sigui ceremonies, has also been found. They also used to organize another special celebration called *bado* to honour Sirius's orbit, which they believed lasted for one year.[34]

Africa was also the birthplace of medicine: Africa produced the world's first physicians, including the first ever father of medicine, Imhotep.[35] Moreover, Africa produced the world's first medical knowledge and literature[36] and influenced and contributed to the work of Hippocrates, the Hippocratic tradition, and the development of medicine in ancient Greece.[37] Greek philosophers and scientists went to Egypt to be educated and returned to Greece with the acquired knowledge. Similarly, many physicians studied in Egypt, including philosopher Pythagorus, who played a major role in early Greek medicine, and Galen, who was an important reference in medicine.[38] Traditional African cultures have always been well-known for treating a number of diseases based on magico-spiritual approaches and plants. The traditional doctor is often an expert psychotherapist who can achieve results with his patients that conventional western psychotherapy is unable to achieve.[39] However, it is worth clarifying for the purposes of this book that African medicine is not limited to spiritual and plant-based healing: examples of obstetric interventions, different types of surgery, bone-setting, removal of bullets, replacement of organs, operations for cataracts, caesarean section, vaccination, anaesthesia, and antisepsis, which were practised for centuries, abound on the African continent. Before the city of Jenne was captured by the Songhay Empire of Sunni Ali in the fifteenth century, it had a university which had high reputation with thousands of teachers who lectured, among other disciplines, in medicine. Jenne was renowned

---

[33] Hunter H Adams, 'African Observers of the Universe: The Sirius Question' in Ivan Van Sertima (ed), *Blacks in Science: Ancient and Modern* (Transaction Publishers 1980) 27.

[34] Ivan Van Sertima (ed), *Blacks in Science: Ancient and Modern* (Transaction Publishers 1991) 12.

[35] Imhotep is regarded as the first physician in history, who lived around 2980 BCE during the reign of Pharaoh Zoser of the third dynasty. Some authors claim that Aesclepios, the Greek god of medicine and present symbol of medicine in the western world, has usurped this position from Imhotep.

[36] The Egyptians were writing medical textbooks as early as 5,000 years ago.

[37] Frederick Newsome, 'Black Contributions to the Early History of Western Medicine' in Ivan Van Sertima (ed), *Blacks in Science: Ancient and Modern* (Transaction Publishers 1980) 127.

[38] ibid 134.

[39] ibid 148.

for its doctors who were able to perform several difficult surgical operations.[40] The contribution of African medicine to the world's pharmacopeia is undeniable: it has produced several effective drugs to treat intestinal parasites, vomiting, skin ulcers, rashes, catarrh, convulsions, tumours, venereal disease, bronchitis, conjunctivitis, urethral stricture, and many other complaints.[41]

In the thirteenth century, while Europe was facing its 'dark ages', the Empire of Mali was enjoying glorious times and exerted control of the famous Timbuktu which was one of the major centres of learning in Africa as it hosted a great university known as 'Sankore'.[42] Timbuktu brought remarkable advancements in human knowledge in areas such as the arts, philosophy, mathematics, and medicine. Its decline started only with the death of Emperor Mansa Musa and the invasion of Tuaregs in 1433. There were also universities and centres of learning in the cities of Gao and Jenne during the Songhay Empire which taught astronomy, mathematics, ethnography, medicine, hygiene, philosophy, logic, diction, rhetoric, and music.

For centuries, the dominant Eurocentric view of the world established that conceptualization of mathematics and physics principles which constitutes the basis for technological sophistication was first developed in Europe.[43] Clearly, the findings illustrated in this book contradict those consolidated Eurocentric ideas that were floated over centuries, claiming that African knowledge has always been intuitive and empirical and was never systematized into abstract concepts. Instead, from these findings it appears evident that Africa was not just a destination of technology but rather the preliminary source of all innovation and creativity. Hence, there is enough evidence to state that the scientific and technological progress currently witnessed worldwide is rooted in the preliminary and fundamental steps that were undertaken in Africa.[44]

b) Africa's technological 'conservatism'

In view of the role played by Africa in the development of the precursor of all technologies—lithic technology—and the great strides made in the development of mathematics, physics, medicine, and astronomy, among others, it is intriguing that the continent did not sustain this pace of progress. It is thus worth investigating the factors that hindered Africa's technological progress.

---

[40] Chu and Skinner (n 1) 91.
[41] Newsome (n 37) 154.
[42] Chu and Skinner (n 1) 88.
[43] Lumpkin (n 18) 100.
[44] The Director General of WIPO said in his remarks on the occasion of the opening of the African Ministerial Intellectual Property Conference held in Dakar, Senegal on 3 November 2015: 'Ultimately, the source of all innovation and creativity is human and Africa is the cradle of humanity, so it is in this sense the origin of all innovation and creativity that characterizes our species as human beings.' Extracts of the speech are available at <www.wipo.int/pressroom/en/articles/2015/article_0014.html> accessed 6 April 2016. Based on the detailed illustration provided in Section C, 'The World's Scientific and Technological Progress' in Subsection C.1a) 'The Contribution of Africa' on 'Inputs into scientific and technological progress', this assertion is not only revealing but seems to be fully substantiated.

The first observation is that the key factor that prompted the pre-industrial societies, such as the ancient Middle East, Greece, Rome, and medieval Europe, to improve productivity and develop faster was to harness non-human energy sources and increase the transformation of linear or reciprocal motion into rotary motion and manufacturing. The reduction of human effort in agriculture, manufacturing, and transportation freed up manpower for the production of goods.[45]

In Africa, the source of energy for agriculture, transport, and manufacturing remained human. Therefore, rudimentary technological implements, if any, used in those areas relied heavily on human energy and linear-reciprocal motion.[46] The question posed is: why did Africa rely on the human factor as the main source of energy to execute those tasks despite the fact that the world had already identified less costly and more efficient solutions? Austin et al point out four main factors in answer to this question: barriers to contact, ecology, demography, and cultural inertia.[47]

The barriers to contact are attributed to the presence of the Sahara Desert to the north and the relatively unindented coastlines and unnavigable rivers in the hinterland. However, the frequent contacts with the outside world that occurred over many centuries discredit this proposition. It seems evident that African societies were exposed to technologies brought by traders and conquerors from other civilizations. However, that did not change the prevailing scenario because Africa nonetheless did not adopt them.[48] Therefore, obstacles posed by the presence of the Sahara, although relevant, were not decisive.

The second potential hindrance to technology transfer is the ecology of the African continent. Indeed, it is an important consideration that many innovations which contributed to greater efficiency in agriculture and transport elsewhere in the world did not work very well in Sub-Saharan Africa. Evidence gathered shows that the nature of the soil and the costs of integrating large, domesticated animals into farming are the factors behind the reluctance to adopt ploughing in Sub-Saharan Africa.[49] The same reasons, connected to the difficulties in raising and using draft animals, made it onerous to adopt wheeled vehicles to facilitate transport and as a result head porterage prevailed.[50]

The third factor is related to demography: reluctance to use animal power in agriculture and transport is linked to the presence of large numbers of slaves around relevant cities. In Africa, it has been always easier and more convenient to

---

[45] Ralph A Austen and Daniel Headrick, 'The Role of Technology in the African Past' (1983) 26 ASR 163, 169.
[46] ibid.
[47] ibid.
[48] ibid 170.
[49] ibid 171.
[50] ibid.

increase productivity by adding slave units to the existing forms of labour, instead of introducing a new technology that could be expensive and risky.[51]

The last factor is cultural and includes economic strategies that were more risk averse instead of focusing on profit maximization. Indeed, risk aversion seems to be embedded in African culture, associating technological innovations with high risk, and thus justifying hesitancy in embracing them.[52] The reality on the ground would seem to endorse this observation. Up until today, fear of the unknown, delaying adoption and adaptation of innovations as a result, appears to be a widespread reaction in the African context.

Low literacy is also a contributing factor in slowing the development or penetration of technology. Indeed, there is a proven strong correlation between literacy and critical-innovative modes of thought.[53] Low rates of literacy prevailing on the continent had a severe impact on the ability to adopt new or foreign technologies.

An unsettling aspect is the 'primacy of politics' which often motivates the African rulers who tend to give preference to political and military rather than economic solutions to social problems. Very often it has resulted in easy to adopt instruments of destruction over the instruments of production. The abundance of land and scarcity of labour justify the preference for military solutions. For example, the Zulu, living in the northern Nguni part of South Africa, were once confronted with a situation of land shortage. Their response to that challenge was not the intensification of production systems but rather the *Zulu mfecane* which was characterized by new forms of state-building, military tactics, and weaponry in order to conquer more land.[54]

The other cultural factor that has reinforced the conservative nature of African technology is the sexual division of labour which expresses itself in two dimensions. The first is connected to the fact that where rotary forms of energy were accepted, the tools were assigned to women, for example the machinery for tending fields and spinning cotton. Men seem to have had a preference for retaining non-rotary forms of energy. The introduction of ploughs or spinning wheels has been resisted partly because they would have meant the involvement of men. This interpretation may be hard to prove although it is undisputable that industries that have adopted rotary systems of production such as the textile industry have traditionally been the domain of women. Cultural conceptions which portray work as synonymous with strength and complexity may have held men back from embracing repetitive forms of work which were instead consigned to women.[55]

---

[51] ibid 172.
[52] ibid 173.
[53] ibid.
[54] Austen argues that the African propensity for political solutions and the investment of technological instruments in the state, especially militarily, would be reinforced by colonial conquest and the various military regimes which have followed independence, ibid 174.
[55] ibid.

Although some of the factors spelt out may be disputed, it is the view adopted in this book that social components can play a pivotal role in the development or adoption of technologies. Strong cultural behaviour is certainly a determinant in the selection of technologies that are acceptable in the local context in view of their compatibility with local perceptions, customs, practices, and beliefs. There is therefore sufficient evidence to conclude that cultural inertia was decisive in setting a low pace in the adoption and improvement of technology and consequently blocking technological progress on the continent.

Comparable patterns are found also in Asia: cultural factors and neo-Confucian philosophy led to the stagnation of the remarkable technological achievements of Chinese society, as will be further elaborated.[56] Indeed, similarly to African society and other civilizations that were far more advanced than Europeans, Chinese society failed to capitalize on the technological advantages that had been achieved in the pre-industrial era. However, European patterns of technological advancement are visibly distinct by emphasizing a more pragmatic approach and inscribing adequate philosophies, epistemologies, and practices that were to prompt individual ownership of ideas, uptake of innovation, and promotion of the IP system.[57]

2. The Contribution of Asia

Asia's contribution to technology, especially stone technology, in the early stages of human evolution was considered negligible.[58] However, as time progressed the situation changed. Specifically focusing on China, from the fourth century BCE to the thirteenth century CE, China's record of invention was unparalleled by any other country in the world. During that lengthy period, China produced impressive inventions and left a major imprint on global technological progress. However, one of the paradoxes surrounding China's invention leadership is that its capacity for inventiveness did not bring wealth and progress to the country. China seems to have failed to move from invention to innovation. Several Chinese inventions were used in their raw form with little improvement. Indeed, in medieval China, the only flourishing activity was agriculture; however, farming innovations usually consisted of incremental, small-scale improvements that could be implemented by individual farmers without the intervention of large-scale research and development efforts.[59] Furthermore, those inventions were not used effectively in the

---

[56] John Alan Lehman, 'Intellectual Property Rights and Chinese Tradition Section: Philosophical Foundations' (2006) 69 JBE 1, 6–8.
[57] Elspeth Whitney, 'Paradise Restored. The Mechanical Arts from Antiquity through the Thirteenth Century' (1990) 80 TAPS 1, 57.
[58] Xing Gao, 'Paleolithic Cultures in China: Uniqueness and Divergence' (2013) 54 CurrAnthrop 358, 368–69.
[59] Ying Lowrey and William J Baumol, 'Rapid Invention, Slow Industrialization, and the Absent Innovative Entrepreneur in Medieval China' (2013) 157 PAPS 1, 15.

productive processes and did not promote technological progress.[60] Therefore, although China also invented printing, the compass, and gunpowder, it was the West that exploited them effectively for its own progress and for the wealth of its own people.

Explanations for China's failure to put inventions to productive use are found in the absence of an innovative entrepreneurial class that could exploit inventions widely and effectively and market their products successfully.[61] The absence of entrepreneurs willing to invest in innovation is attributable to a lack of incentives from the state to enable entrepreneurs to patronize inventors and their endeavours. It was also the lack of incentives for potential entrepreneurs to embrace such a career and the inventors themselves to dedicate their life to inventive activity. Moreover, there were no institutions and rules of law to protect their interests as the patent system was absent and property was subject to confiscation in the name of the emperor.[62] Other factors that are linked to China's limited utilization of its inventions are related to climate, geography, demographic patterns, population growth patterns, political and economic obstacles, the complexity of the written language, and the sense among China's leaders of the superiority of Chinese achievement and knowledge, which led them to resist learning anything from elsewhere.[63] Neo-Confucian thought also shaped perceptions of ownership and exploitation of IP in that, in traditional Chinese thought, to profit from artistic production is considered to be both immoral and low class.[64]

Technological progress was also witnessed in other Asian and Arab countries. However, similarly to what happened to China, it is clear that those achievements did not stand the test of time and did not prevail once confronted with the western wave of technological advancement. Elvin has called this phenomenon the 'high-level equilibrium trap',[65] in which all non-western societies, including those that were extremely advanced technologically such as the medieval Islamic world, the Inca empire, or imperial China up until about 1400 CE, failed to build on their achievements and abruptly stopped their course of technological development.[66]

3. The Contribution of Europe

Western European society was formed between the fifth and the ninth centuries CE as a combination of Roman, Germanic, and Christian influence.[67] Western

---

[60] ibid 2.
[61] ibid; Joseph Garnier, *Traité d'économie politique* (Hard Press Publishing 2019).
[62] Lowrey and Baumol (n 59) 14.
[63] ibid 4.
[64] Lehman (n 56) 6.
[65] George Ovitt, 'The Cultural Context of Western Technology: Early Christian Attitudes toward Manual Labor' (1986) 27 TechnoCult 477, 481–82.
[66] Bray (n 7) 14.
[67] Whitney (n 57) 57.

European society was mainly rural and isolated economically, politically, and culturally from the Eastern Empire.[68] These Germanic kingdoms oriented themselves towards northern Europe rather than the Mediterranean, and developed new forms of political, social, and military organization.[69] During that period, all production was consumed within the household because the economy was based on the production of goods and services to satisfy the needs of families. Those families lived in widely scattered habitations. Technology was rudimentary and only a few itinerant craftsmen would sporadically appear in villages with some implements.[70] However when the population started to increase, villages became larger and closer to each other, and the market grew, providing a better platform for specialized craftsmen to earn a living through the sale of their artefacts.[71] In the first centuries of this millennium, major parts of central and northern Europe developed cities with high population densities. Artisans moved away from the scattered villages and flocked to these towns. Over time, artisans organized themselves. In some towns, different types of artisans were labouring independently, whereas in others, members of a particular craft were dominant, spearheading technological change within their specific field.[72]

Of more crucial importance was scientific and technological development. The increase in population in towns was accompanied by an increase in the number of members of the intellectual elite. Towns were not only centres of craft, trade, and government but also centres of education and science.[73] Indeed, Latin schools, libraries, and universities were established that created a fertile environment for innovation to thrive because European universities became innovators in the natural sciences, replacing foreign science, and hence promoting local technological innovation.[74]

It was during the medieval period in particular that the foundations of the modern western society were laid. Life assumed a new significance in the eleventh and twelfth centuries when society abandoned the idea of 'mere survival in a hostile world' and started to understand and assume that technology and labour could play an active role as transformers of things and objects in order to create a better society.[75] More than in any other earlier society, man invented, used, borrowed, and adapted new devices and machines with a view to transforming society.

---

[68] Indeed Burke sees the formation of European civilization as a reaction against the power constellation of Viking, Islamic, Byzantine, and Mongol civilizations. See Victor Lee Burke, 'The Rise of Europe' (1994) 20 HJSR 1, 9–10.
[69] Whitney (n 57) 4.
[70] Ester Boserup, 'Population and Technology in Preindustrial Europe' (1987) 13 PopulDevRev 691, 694.
[71] ibid.
[72] ibid.
[73] ibid 695.
[74] ibid.
[75] Whitney (n 57) 57.

During this period, members of a craft would organize themselves into a guild or corporation. The guild revolved around families and the apprentices were sons of guild members. Youth not belonging to the family would become servants or unskilled assistants. The guilds were regulated by rules of apprenticeship and passing of professional proficiency tests before establishment as an independent producer.[76] As a result, it was during this period that the modern rigid horse collar was invented and stirrups and iron horseshoes were adopted providing medieval society with a far more efficient use of animal power than had been available in antiquity.[77] The significance of these three inventions and their revolutionary role in modern society is that they supplied Europe with non-human power, at no increase of expense or labour.[78] In this context, White, who in 1940 collected and disseminated information on the contribution of the medieval period to progress said:

> The chief glory of the later Middle Ages was not its cathedrals or its epics or its scholasticism: it was the building for the first time in history of a complex civilization which rested not on the backs of sweating slaves or coolies but primarily on non-human power.[79]

Progressively, the medieval environment became a 'mechanism-minded world' where human beings were surrounded by machines. Western society witnessed the so-called medieval industrial revolution that is the source of modern technological practice.[80] This technological growth was not an isolated phenomenon; it was born in a context of medieval religion, science, philosophy, and attitudes that leant increasingly towards craftsmanship and manual labour. The indifferent, passive, or antagonistic view towards the physical world was replaced by a belief in the moral goodness of labour and technology, a sense of radical separation between man and his natural environment, and the view that man's relationship with nature is properly utilitarian and exploitative.[81] A number of historians, including Chenu, Stock, Ovitt, van Engen, le Goff, and White believe that these new attitudes favourable to technology are embedded in the revised Christian ideas towards work, art, and nature. The classical contemplative idea of nature and antagonism versus the mechanical arts was abandoned; replaced by more concern over technical problems contributing to the formation of a concept of experimental science.[82]

Christian theology played a pivotal role in the new western attitudes towards nature, labour, and wealth and, thus, technology. Based on the biblical injunction

---

[76] Boserup (n 70) 696.
[77] Whitney (n 57) 2.
[78] Lynn White, 'Technology and Invention in the Middle Ages' (1940) 15 Speculum 141, 54.
[79] ibid 156.
[80] Whitney (n 57) 3.
[81] ibid 5.
[82] ibid 7–10.

to rule the earth, Christianity encouraged man to see himself as the master and exploiter of nature. Drawing on the Bible and based on the Book of Genesis 1:27 and 9:6, human beings were understood to be made in the image of God. As a result, man was compared to God in terms of skill in all the arts, including crafts. Work started to be viewed as a spiritual exercise and, as a result, ideals of poverty, humility, and social reform were replaced by the defence of human labour as an aspect of spiritual progress and means of salvation.[83] As a result, monasteries became the first centres of development of technology by matching the practical experience of manual labour and theological notions of work.[84]

Whatever level of contribution to technological receptivity of western society is assigned to Christianity, it is safe to infer that the technological progress witnessed in those states was strongly backed by the full conviction of its people of the legitimacy of human domination over the rest of God's creation and on the sanctity of labour and craftsmanship.

This European cultural context was more favourable to technological progress and enabled Europe to surpass Asia and Africa, the regions that were still struggling with their internal blockages and conservatism.

D. The Birth of Intellectual Property

1. Requirements for the Development of the Intellectual Property System

It is common knowledge that one of the mechanisms that became the main driver of innovation and creativity is the IP system. There is also a consensus in the history of IP that the system was born in Europe and that the first general patent law was passed by the Venetian Senate on 19 March 1474[85] and the first Copyright Act (also known as the Statute of Anne) was passed in Great Britain in 1709.[86] It is also assumed that IP became the main tool for the promotion of innovation and creativity that provided a competitive advantage to western countries. This is evidenced by the fact that as early as the fourteenth century, the King of England went as far

---

[83] ibid 12; Ovitt (n 65) 484.
[84] Whitney (n 57) 12.
[85] It is also accepted that fifty years prior to the Venetian Statute, a prototypical patent was granted by the authorities to Filippo Brunelleschi for a new design of vessel to move loads more cheaply along the Arno River in Florence. However, the Venetian Statute has the virtue of establishing the first formalized patent system. Patents were thus subject to a generalized law, rather than a process of individual petition and grant. Further, grants were based on the applicants' ability to fulfil certain fixed criteria.
[86] Long states: 'The concept of intellectual property has not existed at all times or in all places. [...] I suggest here that although some of the components of the notion of "intellectual property" are evident in antiquity, the fully developed concept first emerges in the medieval period around the 12th or 13th centuries.' She then adds: 'Author's copyright developed after the invention of the printing press, initially in the 16th century', Long (n 7) 848.

as granting foreigners English nationality if they imported new art into England, thus showing the importance of new knowledge and its role in the development of the manufacturing industry.[87] Considering that technological progress was happening all over the world at a broadly similar pace, including in Africa, the question that arises at this stage is: what prompted the birth and development of the IP system only in Europe and why were other regions of the world not embracing it immediately?

The answer to this question is at the root of the requirements for the development of the IP system. There are three prerequisites identified to that end: technological development, a conducive legal and political environment, and a philosophical approach that could conceptualize or allow individual ownership of knowledge.[88] Apparently, Africa's extensively reported technological development was not enough, on its own, to promote the birth of the IP system on the continent. It is therefore worth assessing the other two requisites and establishing where they found fertile soil to enable the IP system to flourish.

2. Political and Legal Framework for the Protection and Promotion of Intellectual Property

a) African context
i) Political environment
The dominant perception is that the manifestations of innovation and creativity occurring on the African continent are not held in high esteem and that ingenuity is viewed as spontaneous and intuitive, hence their outcomes merely accidental. Viewed in this way, it may appear as if local people do not value their creations. However, a closer assessment of arts and inventions that occurred on the African continent and the way these were valued indicates the opposite.

The growth of civilizations promoted the establishment of villages and cities also on the African continent. From as early as 2200–1965 BCE there were iron-processing facilities and blacksmiths' workshops at Ôboui and Gbabibiri, in the northwestern Central African Republic and Gbatoro in Cameroon;[89] between 1,500 and 2,000 years ago populations of Lake Victoria in Tanzania had produced carbon steel; and Engaruka, south of the present Kenya-Tanzania border, was smelting iron before the end of the first millennium.[90] These and many other

---

[87] Pasquale Joseph Federico, 'Origin and Early History of Patents' (1929) 11 J PatOffSoc'y 292, 293.
[88] Christopher May and Susan K Sell, *Intellectual Property Rights: A Critical History* (Lynne Rienner Publishers 2006) 72; Christopher May, 'The Hypocrisy of Forgetfulness: The Contemporary Significance of Early Innovations in Intellectual Property' (2007) 14 RIPE 1, 2.
[89] Zangato and Holl (n 2) 16–17.
[90] Rodney and Davis (n 3) 63.

new settlements became centres of innovation and creativity where craftsmen, woodcarvers, silversmiths, goldsmiths, coppersmiths, blacksmiths, weavers, tanners, dyers, and other ingenuous creators developed and displayed their abilities through their works using the raw materials available in the surrounding natural environment.[91]

In several African kingdoms, high prestige was reserved for artists, as their works could only be afforded and enjoyed by a small minority: members of royal families and people of great importance and prestige.[92] Ivory ornaments, for example, were viewed in some kingdoms and empires as a sign of wealth and power of the royal families. The more ivory stocked, the more power that king could display, and the more artists could flock and compete to make the best works to please him.[93] Through patronage from the kings, artists perfected their art. Among the peoples that mastered the carving of ivory, the Mangbetu, Azande, and Barambo from northeastern Democratic Republic of the Congo deserve special attention in view of the positive role that the royals exerted in promoting creativity. The king of the Mangbetu regarded artists most highly, describing them as the few 'superior men' among the Africans, also recognizing that they were the backbone of his power.[94] It has been said that the king of the Mangbetu would surround himself with persons of aptitude and genius and would protect and support their efforts in order to increase his power.[95] Similar events occurred in Benin, where art forms were also referred to as 'court art' due to the fact that the king maintained the guilds of crafstmen who would only produce art for the king or for religious celebrations or rituals.[96] The Nok and Ife figurines from Nigeria constitute further evidence of the development of art on the continent and show the level of abstraction that artists had reached by 200 BCE and the recognition that they were accorded by the rulers.[97]

The anecdotes narrated demonstrate that the development of art in Africa was a conscious movement backed by the ruling elite. It was therefore deliberately promoted and widely publicized through exhibitions as this also brought prestige to the ruler. The different manifestations of art in Africa are a clear testimony that societies were advanced and could invest time and ingenuity to create abstract works that were not aimed at satisfying solely basic human needs, but even the desires for beauty, pleasure, luxury, and power. Thus, talented people could express their

---

[91] Chu and Skinner (n 1) 69.
[92] Enid Schildkrout and Curtis Keim, 'Mangbetu Ivories: Innovations between 1910 and 1914' (1990) 5 AfriHum 4.
[93] ibid 13.
[94] ibid.
[95] ibid.
[96] Ida Azmi, Spyros Maniatis, and Bankole Sodipo, 'Distinctive Signs and Early Markets: Europe, Africa and Islam' in Alison Firth (ed), *The Prehistory and Development of Intellectual Property Systems* (Sweet & Maxwell 1997) 147.
[97] Chu and Skinner (n 1) 19.

abilities which would in turn satisfy the desires of privileged people and rulers. Similarly, rulers could offer artists special consideration and status in society due to their ability and talent.

In the light of the evidence gathered, it may be inferred that, in addition to the fact that technological progress was happening, the political framework was also conducive to innovation and creativity thriving in early African kingdoms. However, this was not yet enough to enable the birth of a structured IP system on the continent. It is therefore necessary to assess the third requisite: a legal framework for the protection of knowledge and ingenuity in Africa.

ii) Legal framework

The growing importance of the work of creators and innovators in ancient societies slowly prompted the emerging need for protection of the knowledge and the outputs of these skilled and talented people. Protection was therefore crucial for the purposes of regulating the flow of information, maintaining harmony, and settling disputes whenever they occurred. However, the lack of a comprehensive system of protection for the creations and innovative endeavours on the African continent is probably the reason why societies spontaneously developed IP-like protocols to fill the gap.[98] Within the circumscribed and limited context, those mechanisms were effective in protecting the local innovator. Some scholars argue that those systems of protection were sufficient and perhaps more effective than the proposed *sui generis* IP regimes in the current norm-setting mechanisms being discussed and deliberated at the global level.[99] Clearly, that is an overstatement because it is well established that although those systems may have protected knowledge to a certain extent at the local level, they failed to avoid misappropriation when confronted with other sophisticated systems such as those found in Europe.

In Africa, the systems protected knowledge through IP-like protocols, such as the rigid secrecy systems, rituals, and magic.[100] For example, under a secrecy regime, innovative healers only employed their inventions themselves, and benefits accrued to the healer only as long as the medicinal knowledge remained hidden. Furthermore, the treatment could be accompanied by elaborate rituals known only by that specific healer, making his presence compulsory for its effectiveness, even though that treatment was fully disclosed.[101]

In those early societies, technology was simple and the risk of imitators replicating those endeavours based purely on observation was very high. In those

---

[98] Schildkrout and Keim (n 92); WIPO, 'Intellectual Property Needs and Expectations of Traditional Knowledge Holders (WIPO Report on Fact-Finding Missions on Intellectual Property and Traditional Knowledge (1998–1999))' (WIPO 2001) 58 <www.wipo.int/publications/en/details.jsp?id=283> accessed 17 May 2018.
[99] Azmi, Maniatis, and Sodipo (n 96) 99; Schildkrout and Keim (n 92) 60.
[100] Schildkrout and Keim (n 92) 60.
[101] ibid; Mark C Suchman, 'Invention and Ritual: Notes on the Interrelation of Magic and Intellectual Property in Preliterate Societies' (1989) 89 ColumLawRev 1264, 1273.

contexts, the use of magic and taboos in small-scale groups arose in order to protect knowledge that could otherwise be accessed, appropriated, or used by anyone without authorization.[102] Although magic had no concrete function in the physical mechanics of the process, society would construe and view it as being essential to the success of a technique.[103]

Finally, it is worth considering that the effectiveness of the protocols was also guaranteed by customary law enforcement mechanisms. Breaking taboos could result in harsh punishment from the family or clan leader and attract fines payable in local gin, goats, or even a sacrifice. Further disobedience could result in the family being condemned and shunned by the broader family or the community as a whole, including being exiled from it.[104]

Although the efficacy of taboos and magic is undeniable in those contexts, its limitations cannot be overemphasized for various reasons. First, taboos and magic were useful to protect obvious and simple technologies or knowledge that were vulnerable to imitation. Secondly, those mechanisms retained their efficacy in the context of small-scale communities that were bound by the same beliefs and fear of the supernatural. It is therefore not surprising that the efficacy of this proto-legal mechanism crumbled when confronted with people from a different western culture that was less prone to being impressed or influenced by taboos and magic. Thirdly, customary enforcement mechanisms such as the boycott of an infringer's products and banishment from the community only made sense to people belonging to that community and had a very limited effect on strangers.

In conclusion, although the IP-like protocols represented by secrecy regimes, magic, and taboos were successful in setting up a very dissuasive system of monopoly rules to protect knowledge and invention which commanded a lot of respect and full compliance in the community, these did not go beyond the limits of that same community. The system failed to evolve into a more systematized body of rules capable of challenging the test of territory, time, and cultures. Further, rather than promoting innovation for the benefit of the entire community, IP-like protocols based on the 'mythology of risk' froze opportunities for the sharing of technical information, blocked imitation, and stifled innovation on the continent. It is therefore safe to infer that although the existing kingdoms and empires recognized and promoted the work of creators and innovators, they failed to develop mechanisms of protection of their knowledge and output, and, more importantly, to develop a more abstract policy, or legal or institutional framework for the protection of knowledge of their citizens.[105]

---

[102] Suchman (n 101) 1272.
[103] ibid.
[104] Paul Kuruk, 'Protecting Folklore Under Modern Intellectual Property Regimes: A Reappraisal of the Tensions Between Individual and Communal Rights in Africa and the United States' (1999) 48 AmUniLawRev 786 <https://digitalcommons.wcl.american.edu/aulr/vol48/iss4/2> accessed 17 May 2018.
[105] Suchman (n 101) 1279.

The obvious finding, therefore, is that although innovation was happening in Africa and knowledge, creativity, and ingenuity were valued by the rulers, the latter frustrated their innovators and creators by failing to establish an effective and objective system of protection of their endeavours. Regrettably, the African continent is yet to overcome the inability to master the relevance of the IP system to its own development and to establish a system that caters for its own interests as will be illustrated throughout this book.

iii) Individual ownership of knowledge

There is in fact no doubt that IP systems need an environment where technological development is vibrant and where the political authorities provide the necessary legal and political framework to promote it. However, as already discussed, these two requisites were not sufficient for IP rights to thrive. The decisive factor was the trend towards distinguishing intangible property from material products and, more importantly, to attaching commercial value to that intangible property which was further attributed to an identifiable individual that signalled the birth of IP rights. The culture of recognition of the individual as sovereign creator of knowledge and thereby his legitimacy in appropriating knowledge is the building block of an IP system. This validation of individuals as legitimate holders of knowledge should be assessed in Africa, Asia, and Europe with a view to determining its implications in the development of the IP system.

Failure to develop an IP system in Africa is attributed largely to the lack of knowledge appropriation by single individuals. Contrary to the western-style system of innovation and its corresponding IP system that prioritizes the development of the market economy through individual commercial gains, innovation occurring in Africa was communal, aiming at satisfying the social well-being of society as a whole.[106] Although Africa witnessed the establishment of guilds to regulate the activities of innovators and creators and to cater for their welfare, the recognition of individual ownership of knowledge and its protection as such did not occur in those African guilds. To ensure that the knowledge was kept within the castes or guilds and would guarantee privileges to those with specific skills, the knowledge was protected through sophisticated communal regimes of ownership and control of invention in those ancient societies.[107] This protection was an efficient way to ensure that innovators and artists in the communities managed to derive a fair and equitable share of the benefits arising from the use of their innovations within the community. Those benefits consisted of the exchange of their services for monetary payments, commodities (food, trinkets, etc); event and service-based

---

[106] WIPO (n 98) 220.
[107] Poku Adusei, 'Trajectories of Patent-Related Negotiations Affecting Pharmaceuticals and the Politics of Exclusion in Sub-Saharan Africa' (2010) 24 UGLJ 38.

benefits (improved access to infrastructure, dedicated feasts and celebrations); and non-monetary benefits such as social prestige and political influence.[108]

The birth of guilds on the African continent was necessitated by the challenges faced by local creators in terms of the regulation of their activities and to cater for their material and spiritual needs. As a result, the guilds operating in Africa were not established with proprietary tendencies and did not prioritize individual economic rewards. Members of the African guilds would deem themselves to be satisfied when the community provided their members with an enhanced reputation and they were able to meet their basic necessities such as food and clothing. However, knowledge continued to be protected through secrecy, as a collective good within the guild, without specifically recognizing and exalting the individual behind the innovation. Therefore, contrary to what happened in other geographical areas, the recognition of individual ownership of knowledge and its protection as such did not occur in African guilds. Although sharing and collectivism of knowledge is a sublime ideal and is morally commendable, it is observed that this deprived African innovators and creators of the necessary motivation to consistently pursue their talent and rendered them less competitive when compared to those operating in other systems developed in other regions, such as Europe. Therefore, it is submitted that sidelining innovators and creators by giving prominence to the community kept them continuously at subsistence level and subjected them to acceptance, as a reward for their talent, only of what was required for the satisfaction of their basic needs.

As a result, IP did not find a fertile environment to develop and the generosity and culture of sharing among African peoples has consistently held the continent back in its quest for the development of an IP system geared towards its social, economic, and technological progress.[109]

---

[108] WIPO (n 98) 62.

[109] Notwithstanding the millennia that have elapsed, in many parts of Africa proprietary attitudes towards intangible assets and attaching commercial value to them are concepts that populations are still struggling to absorb:

> Generally in Kenya, people have a culture of giving and sharing. This culture is so embedded in the community that it is almost absurd to ask people to pay for using other people's works. Everyone's assumption is that creators must create works for the communities to use and benefit from [...] The culture of sharing and free use is so embedded in our societies that licensing and payment is viewed as a foreign concept. We share what we create with the community and hope that in return the community somehow pays back what you have given.

See Sharon Chahale, 'Cultural Barriers to the Introduction of Western Model Collective Management Systems in Kenya' in Anand Nair, Claudio Tamburrino, and Angelica Tavella (eds), *Master of Laws in Intellectual Property—Collection of Research Papers* (Edizione Scientifiche Italiane 2013) 376. Informants of the WIPO fact-finding missions conducted in 1998–99 viewed the IP system as a modern reincarnation of European colonialism. They indicated that the notions of 'property rights' and 'ownership' are foreign to indigenous and local communities, WIPO (n 98) 217.

b) The Asian context
i) Political environment

Asia's dominance in knowledge was evident from the fourth century BCE to the thirteenth century CE, with China leading the world and providing impressive inventions that left a major imprint upon global technological progress.[110] However, China's technological progress was blocked during the Ming Dynasty (1368–1644 CE). The Ming Dynasty's approach of isolating China from the West following the earlier introduction of neo-Confucianism by the Sung Dynasty blocked the inventiveness of the Chinese people and discouraged the advancement of technology in the country for many years.[111] Therefore, the impressive technological progress that was achieved up until the fourteenth century was trapped by the rigid 'bureaucratic feudalism' of the neo-Confucian tradition.

Neo-Confucian thought shaped perceptions of ownership and impeded the birth of IP in China. Under the neo-Confucian culture, it was deemed immoral, unethical, and low class to profit from artistic production. Furthermore, the neo-Confucian and Taoist doctrines discouraged individualism and promoted the sharing of property, including inventions, discoveries, and creative works. In view of this, inventors and creators could expect only public recognition and endowments given by the ruling king or emperor as a reward for their hard work and successful intellectual achievements.[112] This intellectual and cultural dissonance rooted in neo-Confucian and Taoist thought not only impeded the birth of IP in China but was also the main cause of the challenges in the development of the IP system and the enforcement of IP rights that China witnessed up until the early 2000s.[113]

ii) Legal framework

As a result of the clear and unwavering position of the Chinese state, there was no nascent group of capitalists able to further their own opportunities for accumulation and to promote innovation.[114] The absence of entrepreneurs willing to invest in innovation, due to lack of incentives from the state, deprived inventors of the support necessary to pursue technological progress. Furthermore, China did not provide for institutions and laws to protect their interests. Indeed, there was no patent system and property was subject to confiscation at any time and for any reason in the name of the emperor.[115] In such circumstances, there could be no

---

[110] Lowrey and Baumol (n 59) 1.
[111] Ovitt (n 65) 481–82.
[112] Jennifer Wai-Shing Maguire, 'Progressive IP Reform in the Middle Kingdom: An Overview of the Past, Present, and Future of Chinese Intellectual Property Law' (2012) 46 IntLawyer 893, 896.
[113] Lehman (n 56) 8.
[114] ibid.
[115] Lowrey and Baumol (n 59) 14.

expectation that any kind of individually owned IP rights would flourish as this was not an inherent part of the system in that specific context.

One of the industrial sectors that did progress well in China was the printing industry. However, even this industry did not prompt the need for an adequate legal framework that could promote creativity and innovation. All sets of rules established were more concerned with securing the publishing monopoly of the state: the Tang Dynasty (618–907 CE) adopted a legal code prohibiting the transcription and distribution of a wide range of literature in order to protect the emperor's prerogatives and interests. For example, the first known ordinance regulating publication issued by Emperor Wen-Tsing in 835 CE pertained more to the prohibition of private publication of almanacs; and the Sung Dynasty (960–1179 CE) adopted an extensive regulatory apparatus and official government printing houses in the major cities also with a view to having full control of the printing industry.[116] Although a few private printing houses appeared, printing works depended on exclusive privileges granted to them by imperial officials. Therefore there was no right to print, but only a limited privilege granted by the government.[117]

In the copyright domain, it is worth emphasizing further that China tolerated copying as it was considered a legitimate way of learning, and also served to maintain the tradition of isolationism and distrust of outsiders. This approach was another factor that hampered the development of an IP system.[118] For a long time, China jeopardized its own progress due to inadequate policies, neo-Confucian and Taoist culture, and the selfishness of the ruling class. As a result of the prevailing situation, although China progressed well in some fields, paradoxically it missed the opportunity to establish the building blocks for the development of a systematic legal framework capable of promoting innovation and creativity, namely through the development of an IP system.[119]

iii) Individual ownership of knowledge

A glimpse into other great civilizations of the premodern world—Chinese, Islamic, Jewish, Christian, and Greek—reveals that human ownership of their ideas or expressions was also absent.[120] In the Chinese Confucian tradition, knowledge and writing were not subject to commerce.[121] Authors practised their craft for the moral improvement of themselves and others, and therefore reputation— especially the esteem of future generations—was their only reward.[122] The Chinese were also engaged in the development of technological innovations similar to

---

[116] Carla Hesse, 'The Rise of Intellectual Property, 700 B.C.–A.D. 2000: An Idea in the Balance' (2002) 131 Daedalus 26, 29.
[117] ibid.
[118] Maguire (n 112) 897.
[119] May (n 88) 16.
[120] Hesse (n 116) 27.
[121] ibid.
[122] ibid.

those in European countries and developed strong ideas related to the need to own knowledge. However, in the Chinese case those same innovations failed to inspire the need to develop an IP system.[123] Similarly, Budde-Sung stresses that the recognition of IP as something that belongs to someone and requires protection may be a function of individualism.[124] Therefore, a country with a more collectivist approach, such as Japan, did not pass any IP laws until 1885. Buddhist countries, influenced by Confucian ethics, also perceive the imitation of creative endeavour as praise for their work, and that this imitation may help the society overall to learn and improve.[125]

The Islamic world also ignored any concept of IP for many centuries. All knowledge was thought to come from God. The Koran was the single greatest scripture from which all other knowledge was derived, and any text that embodied the word of Allah could not belong to anyone.[126]

The Judeo-Christian tradition also viewed knowledge as a public good to be shared and not to be possessed by anyone. Accordingly, Moses was the one who received the law from Yahweh and freely transmitted it to the people chosen to hear it. Knowledge was therefore seen as a divine gift and consequently it could not be sold. During the medieval period, theologians admonished university professors, lawyers, judges, and medical doctors not to charge fees for their services although they could receive gifts in gratitude for the wisdom they imparted.[127] In Greece, although the sophists were the first group to earn significant rewards through their freelance teaching activities, they never regarded knowledge or information itself as a commodity that could be owned.[128]

In conclusion, the requirement of individual ownership of knowledge was not nurtured in all the great civilizations of the premodern world.[129] Apparently the absence of the three concurrent requisites for the development of an IP system, namely technological progress, a legal and policy framework, and the legitimacy of private ownership of knowledge appear to be a common denominator of the civilizations in Africa and Asia. This explains why the IP system did not find a fertile environment to put down roots and thrive in such circumstances despite the fact that innovation and creativity were taking place.

---

[123] May (n 88) 16.
[124] Amanda Budde-Sung, 'The Invisible Meets the Intangible: Culture's Impact on Intellectual Property Protection' (2013) 117 J BusEthics 345, 347.
[125] ibid 346.
[126] Hesse (n 116) 27.
[127] ibid 28.
[128] May (n 88) 45.
[129] ibid 17.

c) The European context
i) Political environment

Although technological development was happening in many areas in Europe, especially within guilds, up until the medieval period the continent was still divided into many small independent states.[130] That environment was not yet conducive to granting and protecting IP rights. In the circumstances, inventors and manufacturers still relied more on maintaining the secrecy of their processes rather than resorting to patents.[131] However, the political environment related to the protection of knowledge and technology in Europe witnessed evolution and change throughout this period. Indeed, in order to stimulate innovation and inventiveness, medieval monarchs adopted deliberate policies such as: i) granting monopolies and privileges to innovators to induce them to embark upon hazardous undertakings; ii) granting special franchises and privileges, including citizenship, to foreigners who introduced new arts from abroad and settled in the kingdom; and iii) attracting foreign skilled artisans to the kingdom for the furtherance of national industry.[132] Manufacturing and trade benefited tremendously from these royal initiatives to grant monopolies and special privileges in order to promote public interest. Examples of such measures are found in England, Germany, and the Republic of Venice (now part of Italy), and the development of some industries such as the manufacture of salt, armoury, shipping, printing, glass, and clothing is attributed to those policies.[133]

By the fifteenth century, awareness of the importance of the privileges to be accorded to some manufacturers was consolidated. Increasingly, guilds were granted easy access to the government of Venice, enabling them to make their needs known directly and to lobby for solutions. The patent law of 1474 was enacted as a result of pressures exerted by the influential guilds on the Venetian Senate in order to strengthen their control over knowledge.[134]

The developments described lead one to assume with confidence that the transition from mere acknowledgement of innovators to the granting of privileges, which further evolved into rights, was prompted by a political class sensitive to the imperatives of technological progress. Although self-interest and especially material gain were pursued by the European ruling class, the desire to realize public interest allowed it to concede space for innovation to thrive. A ruling class that was inclined to promote technological progress and the ability to enact policies capable of paving the way for a system of recognition of private rights to knowledge and its protection through a legal system that promotes rights to intangible goods was instrumental in developing an enabling environment for innovation. This was

---
[130] Federico (n 87) 295.
[131] ibid.
[132] ibid 294.
[133] ibid 293.
[134] May (n 88) 6.

contrary to what was witnessed in China, where the self-interest of the ruling dynasties prevailed over the noble objective of stimulating technological progress for the benefit of the nation, therefore stifling innovation and progress.

ii) Legal framework

Governments in Europe took action to promote innovation in a more vigorous manner by developing legal instruments for the protection of inventions. For example, as early as 1297 the Great Council of Venice passed a decree concerning the manufacture and sale of medicines that gave physicians monopoly rights over their secrets, provided they used high-quality ingredients.[135] The law signalled the desire by the local government to promote invention and thus the innovation and development of the economy of the city using the IP system although only at its infant stage. In analysing that law, it emerges that, already by that time, it contained some of the main features of the modern IP system, such as the obligation to disclose the content of the invention, the underlying right of the inventor to compensation, the limited duration of the rights, the enforcement of IP rights, and the concept of expropriation or, according to modern terminology, the compulsory licence.

In England, some policies and statutes are identified even before the first patent law was enacted in Venice in 1474. These are: the policy of encouraging the importation of new arts from abroad enacted in 1326; and the statute of 1337 that established that all cloth workers from other states would be given special franchises and privileges if they settled in England and practised and taught their arts.[136] In 1323, the Council of Venice agreed to support, with the sum of 80 ducats, the 'inzenerius molendinorum' of one Joannes Teutonicus in his experiments as he made machines able to meet the needs of Venice, which is clearly one of the first signs of financial support for innovation by a state.[137]

As a result of these and other laws and statutes enacted in Europe, many privileges were granted in diverse fields of technology. Federico cautions us that those initial grants were only aimed at the introduction and establishment of new industry. No formal written disclosure was required; patents for improvements were of doubtful validity; and the patent was invalid if a prior use could be shown.[138] Therefore, what can be inferred is that the main consideration of initial patents during the medieval period was not the reward of inventors, but a much higher public interest in the introduction and establishment of a new industry. Furthermore, the requirement of disclosure of the patent through a written document was not fundamental at that stage. The rights conferred enjoyed high consideration and protection and could be enforced across Europe. For example, false

---

[135] Long (n 7) 876.
[136] Federico (n 87) 293.
[137] Long (n 7) 877.
[138] Federico (n 87) 305.

marks on cloth, gold, or silver were treated as counterfeiting currency, punishable by death.[139] Accordingly, in Europe, by the end of the medieval period, the policy and legal environment was ripe for the birth of the i system and governments took concrete steps to promote and protect knowledge.

iii) Individual ownership of knowledge

The concept of IP was developed based on the proprietary attitudes towards craft knowledge that were matured within the medieval system of guilds in Europe.[140] The system of guilds, which consisted of organizations of artisans involved in specific crafts or technologies, occurred worldwide in ancient times, as highlighted in the case of the African continent. During the rule of Emperor Augustus (27 BCE–14 CE) in Roman territories, guilds were commonplace. But, by the third century CE the guilds were transformed from voluntary associations with government approval into 'compulsory public service corporations entirely controlled by the state'.[141] Contrary to the later format of the guilds, these organizations were not designed primarily to train apprentices, set standards, or protect craft secrets, but were instead aimed at attending to the cult of a deity, or to provide burial for poor members, or to meet for social purposes at feasts and banquets.[142] Therefore, the original format of the craft guilds did not involve economic interests but religious, social, and funerary functions. It is worth emphasizing at this stage that those who possessed unique skills of craft or invention were humble in their possession of them without expecting any special recognition or reward. Their concern was technological advancement and to find solutions to the challenges faced by the society without any ambition of owning them.[143] It is for this reason that creators such as Theophilus, who wrote a twelfth-century treatise on painting, glass, and metalwork, *De diversis artibus*, abrogated authorship and wrote under pseudonyms. Theophilus advocated the open sharing of human skill and knowledge for the dual purposes of human advancement and the glory of God.[144]

During this period, talented people who were actively bringing improvements to many areas of knowledge would satisfy themselves with food, safety, and being excused from hard labour as well as other symbolic rewards dispensed by the dominant classes.[145] Therefore, it should be appreciated that the building blocks of the western IP system were initially laid when the guild's approach was still based on collective ownership of knowledge focusing on religious, social, and funerary functions, similarly to what was happening in other parts of the world.

---

[139] May (n 88) 9–10.
[140] Long (n 7) 870.
[141] ibid 862.
[142] ibid.
[143] ibid.
[144] ibid 868.
[145] ibid 869.

THE BIRTH OF INTELLECTUAL PROPERTY   47

This scenario changed radically in Europe when the medieval craft guilds became more urbanized. The urban environment in which the medieval guilds operated fostered the development of proprietary attitudes towards craft knowledge.[146] The Italian cities (and specifically Venice) played a major role in the development of the proprietary attitudes that fuelled the birth of the IP system as it is known today.[147] More decisively, it became clear that craft knowledge was intangible property with commercial value, and therefore guilds began to recognize that individual members might have an exclusive right to certain knowledge. For example, the general articles for the governance of the Genoese silk manufacturers of 1432 provided for the protection of patterns or figures designed by one individual. Similar provisions existed in the Florentine Woollen Guild by 1474 to protect the designs of certain fabricators.[148] Furthermore, some individuals broke from their guilds and took guild knowledge with them, later claiming the same as individual knowledge in other jurisdictions. Philosophically, it appears critical that it was during the Renaissance in Italy that a more individualized notion of creativity aimed at promoting the reward of individuals for their innovative or creative effort was crystallized.

Guilds learned to value their craft secrets and to derive associated advantages from society. This was achieved primarily by establishing symbols to mark their goods: the symbol would ascertain the origin and due inspection of goods to guarantee that those goods were approved and met defined standards. Goods originating from a specific guild incorporated an added value that could differentiate them from non-guild goods whose standards could not be ascertained. Long finds in this reasoning the precursor of patents, because it shows how craft processes slowly became intangible property and their owners realized the commercial value of such processes.[149]

It is therefore widely accepted that the basic ingredients and conditions for the development of the IP system were crafted in the guilds in Europe and only subsequently was the framework adopted by the juridical authorities.[150] Most importantly, although guilds were present in many parts of the world, the difference that existed between the western-style guilds and those guilds located in other regions, such as the African guilds, arose when the European guilds identified craft knowledge as intangible property, attaching to it commercial value that signalled the birth of IP rights. Therefore, the narrative of the birth of the IP system points to the development of knowledge ownership, which began to attract formal protection in Europe from the fifteenth century onwards with regard to material inventions, and later with regard to writings, in the sixteenth century.[151] Indeed, as highlighted in

[146] ibid 870.
[147] ibid 875.
[148] May and Sell (n 88) 51.
[149] Long (n 7) 874.
[150] May and Sell (n 88) 871.
[151] Long (n 7) 847.

the introduction of Section D 'The Birth of Intellectual Property', the first general patent law was passed by the Venetian Senate on 19 March 1474[152] and the first Copyright Act (also known as the Statute of Anne) in Great Britain in 1709.[153]

The assumption is therefore that the origin of the IP system as it is known today can be traced to Europe. Unfortunately, all other knowledge protection regimes developed in other civilizations, even those that predate the European IP system, such as African IP-like protocols, did not evolve into sophisticated and modern legal systems. Consequently, those regimes did not prevail when applied beyond their original territory or community and were easily sidelined by more systematic frameworks from western countries.[154] In view of that, the western-style IP system became the new and sole mode of promoting and recognizing human ingenuity.

E. Conclusion

The scientific and technological progress currently witnessed worldwide is rooted in the preliminary and fundamental steps that were undertaken in Africa.[155] However, that progress abruptly came to a standstill in Africa and Asia in what some historians have designated as blockages, brakes, or traps—specifically indicated as 'high-level equilibrium traps'—which are largely linked to the cultural context of the respective civilizations.[156] While that was happening, Europe was devising the most powerful incentive to innovation: the IP system.

Apparently, an IP system only thrives in a context where three prerequisites are simultaneously present, as observed by May, namely: technological progress, a legal and political framework, and the philosophical conceptualization of the individual as sovereign knowledge producer and therefore legitimate owner thereof.[157] Those three prerequisites were present contemporaneously only in Europe during the Renaissance period, prompting the birth of the IP system.

Africa is characterized as having failed dismally to attach value to the immense craft and inventive knowledge that was thriving on the continent. Guilds and rulers failed to depart from the model that protected innovation as a collective guild knowledge and to recognize appropriation of knowledge by individuals, thereby

---

[152] 'Thus the law required that each person who invents a new and ingenious device, not made before in the Venetian domain, must give notice of it at the *Provveditori di Comun*. No one within the Venetian domain could make a similar device for ten years. If anyone violated the law, the "author and inventor" could report it to city officials. The infringer could be required to pay 100 ducats and the device might immediately be destroyed. However, the Venetian government could itself take and use it for its own needs as long as others did not', ibid 878.

[153] ibid 848.

[154] Chidi Oguamanam, 'Local Knowledge as Trapped Knowledge: Intellectual Property, Culture, Power and Politics' (2008) 11 JWIP 29, 33.

[155] Sertima (n 34) 4.

[156] Bray (n 7) 14.

[157] May (n 88) 2.

constituting a serious impediment to innovators and creators having access to better rewards and incentives. Although collectivism and a culture of sharing are hailed as virtues on the continent, unfortunately they are also responsible for keeping the continent less competitive and Africa is therefore struggling in its quest to use the IP system for the promotion of its social, economic, and technological progress.

3
Dynamics of the Extension of the Western-Style IP System into the African Continent

A. Introduction

In January 1898, the British Secretary of State for the Colonies consulted the then Acting Governor of the Gold Coast (known today as Ghana) regarding the procedure for obtaining patent protection in that colony. Apparently, there was no patent legislation in that territory at the time and such a statute had to be specifically enacted, triggered by that demand. The legislation was enacted in the form of a Colonial Patent Ordinance—the Gold Coast Patent Ordinance no 1 of 1899.[1]

This incident was not unique to Ghana. Similar episodes that occurred in various African countries confirm the same pattern in the process of introduction of the intellectual property (IP) systems on the continent. In a nutshell, colonization provided a unique opportunity for the extension of the IP systems developed and implemented in Europe in the newly discovered territories of the African continent. This was prompted by the need to protect IP rights of western citizens and companies operating in Africa, since adequate mechanisms for the protection of their intellectual endeavours were lacking on the continent. Hence, the real motivation for the introduction of western-style IP systems was connected to the desire to extend privileges and benefits to European citizens who were spread throughout the territory of Africa using a system that was more familiar and suitable for them.

The imposition of western IP systems in Africa has been highly criticized for replacing local knowledge and innovation protection systems, thereby denying African countries a chance to develop. For example, Okediji asserts that Europeans exploited the weaknesses and vulnerability of the African systems to justify the use of the European IP laws.[2] Penrose indicates that such blind extension of IP into unknown territories, guided by a 'one size fits all' belief on the part of the colonial masters, further contributed to the marginalization of local knowledge and its protection mechanisms.[3] Nevertheless, the 'blame game' to explain

---

[1] Samuel O Manteaw, 'Patent Law in Ghana: Proposals for Change' (2010) 24 UGLJ 114–15.
[2] Ruth L Okediji, 'Africa and the Global Intellectual Property System: Beyond the Agency Model' (2004) 12 AYIL 207, 216.
[3] Edith Tilton Penrose, *The Economics of the International Patent System* (Johns Hopkins Press 1951) 53.

Innovation in Africa. Fernando dos Santos, Oxford University Press. © Fernando dos Santos 2024.
DOI: 10.1093/oso/9780192857309.003.0003

underdevelopment of Africa due to the negative impact of the implementation of the western-style IP systems seems untenable when 'non-colonized' countries of Africa, namely Liberia and Ethiopia, are considered. For example, Ethiopia was guided by a limited vision of the ruling class which prevented society from developing traditions and institutions which would have protected the property of merchants and craftsmen, hence fostering individual creativity.[4] The dawn of political and economic freedom in the period between 1960 and 1970 did not improve the scenario on the continent, as African countries persisted in ignoring the crucial role that IP can play in promoting their development. The industrial policies developed after independence are conspicuous in mainstreaming IP into the development policies. Therefore, this approach ought to be reviewed, and deep introspection be undertaken in order to identify the endogenous factors that are impeding the technological development of Africa such as the lack of use of the IP system and provide solutions.

This chapter invokes historical methodologies and analysis of international IP legal instruments to assess how knowledge and innovation were protected in Africa, how a western-style IP legal system was transposed into the African continent, and the impact thereof on Africa's technological development, while also scrutinizing IP systems in the African countries that were never colonized and their correlating technological development. The aim is to determine whether imposition of IP systems is indeed the culprit in Africa's lack of technological development or whether there are other factors.

The chapter is structured as follows: the introductory section contextualizes the topic by describing succinctly the system that was in use in Africa for protecting knowledge and innovation prior to the arrival of Europeans, including its limitations. The second section analyses the pre-colonial extension of the western-style IP systems in Africa, highlighting the reasons and the rationale behind it and the modalities and the structures that were established to enable the system to work in the new territories of Africa; the impact of the implementation of this alien system on indigenous knowledge is then described. Subsequently, the non-colonized states of Africa (Ethiopia and Liberia) are analysed with a view to establishing whether in the absence of the western-style model, the continent could have moved onto a different path of technological progress, and conclusions are then drawn. Post-independence implementation of IP systems in selected countries are further considered. To conclude, some best practices in mainstreaming IP policies into the national development programmes are illustrated, with the aim of enabling other countries to emulate them.

---

[4] Merid Wolde Aregay, 'Society and Technology in Ethiopia, 1500–1800' (1984) 17 IJES 127, 133–43.

B. Pre-colonial Extension of the Intellectual Property System to Africa

1. Protection of Knowledge and Innovation in Pre-colonial Africa

The growing importance of the work of creators and innovators in ancient societies in Africa called for protection of the knowledge and works of its skilled and talented people. However, due to lack of a comprehensive system of protection of the creations and innovative endeavours on the African continent, societies spontaneously developed IP-like protocols to fill the gap.[5] Apart from protection, the system was useful for regulating the flow of information, maintaining harmony, and settling disputes whenever they occurred.

Although those protocols were successful in setting up a very dissuasive system of rules to protect local knowledge and innovation, which commanded a lot of respect and full compliance in the community, they did not transcend the limits of that same community. The system therefore failed to evolve into a more systematic body of rules capable of challenging the tests of territory, time, and cultures. Further, rather than promoting innovation for the benefit of the entire community, IP-like protocols based on the 'mythology of risk' froze opportunities for the sharing of technical information, blocked imitation, and stifled innovation on the continent. Therefore, although the then African kingdoms and empires recognized and promoted the work of creators and innovators, they failed to develop sustainable mechanisms of protection of their knowledge and output, and, more importantly, to initiate a more abstract body of rules and systems capable of protecting the knowledge of their citizens and to foster innovation.[6]

Absence of a solid mechanism for the protection of knowledge was a gap also felt by the Europeans upon arrival but was concomitantly an opportunity that was effectively exploited subsequently to install their own model, which ultimately benefited them.

---

[5] WIPO, 'Intellectual Property Needs and Expectations of Traditional Knowledge Holders (WIPO Report on Fact-Finding Missions on Intellectual Property and Traditional Knowledge (1998–1999))' (WIPO 2001) 58 <www.wipo.int/publications/en/details.jsp?id=283> accessed 17 May 2018; Enid Schildkrout and Curtis Keim, 'Mangbetu Ivories: Innovations between 1910 and 1914' (1990) 5 AfriHum 60.

[6] Mark C Suchman, 'Invention and Ritual: Notes on the Interrelation of Magic and Intellectual Property in Preliterate Societies' (1989) 89 ColumLawRev 1264, 1279; Paul Kuruk, 'Protecting Folklore Under Modern Intellectual Property Regimes: A Reappraisal of the Tensions Between Individual and Communal Rights in Africa and the United States' (1999) 48 AmUnivLawRev 786.

2. Rationale of the Introduction of Western-Style Intellectual Property Systems in Africa

By the eighteenth century, European civilization was achieving the full maturity exemplified in the Scientific, Industrial, and French Revolutions.[7] The Industrial Revolution increased the need for raw materials such as gold, diamonds, palm oil, and other commodities to satisfy the needs of the growing European industry. This increasing demand for raw materials and markets, together with other geopolitical pressures, forced European states to engage in a more vigorous approach to imperialism and colonialism.

The Berlin Conference was held from 15 November 1884 to 16 February 1885 and became the new landmark in Africa's history as the European powers ruled on the Scramble for Africa and a new cartographic constitution and configuration of the continent.[8] One of the resolutions of the Berlin Conference concerned the principle of effective control, which became one of the major criteria in proving possession of African territory by European powers. The effective territorial occupation of African communities by European powers was achieved through military action and settlement but in some cases could be proved by treaties and agreements signed with local rulers.[9] By 1914, the whole of Africa had been brought under colonial rule, except for Liberia and Ethiopia.[10] The effective occupation of the land by European powers paved the way for the phenomenon of colonialism on the African continent. As a result of this process, the IP systems developed and implemented in Europe were extended to the newly discovered territories, especially on the African continent.[11] The reason for this is that the western powers did not find in Africa and elsewhere adequate mechanisms to protect their rights and they exploited the weaknesses and vulnerability of the African systems to justify the use of European IP laws where suitable and necessary.[12] Therefore, the circumstances and

---

[7] John D Fage and William Tordoff, *A History of Africa* (4th edn, Routledge 2001) 326–28.
[8] Sabelo J Ndlovu-Gatsheni, 'Genealogies of Coloniality and Implications for Africa's Development' (2015) 40 AfrDev 13, 26.
[9] Ndlovu-Gatsheni remarks that:

> Use of treaties and concessions bearing the signatures of African kings and chiefs must not be taken to mean that African leaders consented to colonisation. The treaties were obtained fraudulently through trickery, chicanery and outright lying by European negotiators and agents. One of the examples is the Rudd Concession of 1888 that was claimed to have been signed between the agents of the British South Africa Company (BSAC) and King Lobengula Khumalo, the last leader of the Ndebele Kingdom in southern Africa in the immediate post-Berlin Conference period. What obtained later is that the pre-literate Ndebele king had not understood the terms of the treaty that were written in English and there was a difference between what was shared with the Ndebele king verbally and what was contained in the written treaty. When the true facts of the Rudd Concession were later understood by the Ndebele king, he immediately and vehemently repudiated it (ibid 27).

[10] ibid.
[11] Okediji (n 2) 216.
[12] ibid.

timing of the introduction of the IP system in Africa did not follow the European pattern. Indeed, in analysing the international IP legal framework, the evidence suggests that the Berne Convention of 1886 was neither designed to advance indigenous arts and culture in the African territories (as this was not taken into account in the negotiation of the Convention) nor to benefit the local writers as the majority of the indigenous populations could hardly read or write.[13] It was, instead, the desire to protect the literary works produced by European authors through harmonized laws and especially the need to curb the rampant piracy in Europe and America that necessitated an international copyright system. With regard to the Paris Convention on Industrial Property of 1883, it is obvious that it was designed to protect inventions exploited beyond national borders, with the ultimate objective of advancing the self-interest of the colonial powers for their own national technological progress.

Penrose traces the source of the current patent system on the African continent to the series of conferences held in Vienna and Paris in 1878, 1880, and 1883, respectively, on the establishment of an international patent system, including a mechanism to extend this to the colonies, possessions, or protectorates.[14] Clearly, such a blind extension in unknown territories was guided by a 'one size fits all' belief by the colonial masters and for the need to provide for western interests in the conquered territories. The narrative of the historical development of patent law in Ghana illustrates clearly the mismatch between the African state's interests, including its developmental objectives, and the alien rationale of the introduction of IP laws in African territories. Thus, in January 1898, the British Secretary of State for the Colonies sent a dispatch to the then acting governor of the Gold Coast (known today as Ghana) enquiring about the procedure for obtaining patent protection in the Gold Coast colony, apparently after receiving an enquiry from Messrs H & W Pataky, patent agents in the United Kingdom (UK). Since there was no patent legislation in the colony, the Secretary of State subsequently asked the Board of Trade for advice on the need for a patent law for the Gold Coast. Ultimately, the legislation was enacted for the Gold Coast in the form of the Gold Coast Patent Ordinance no 1 of 1899.[15]

The circumstances described here demonstrate that it was not the desire to encourage either indigenous inventive activity, innovation, or technology transfer that prompted the adoption of the ordinance. Rather, the real motivation for the adoption of that legislation was the development of the gold mining industry, specifically in the district of Tarkwa in the then Gold Coast and the resulting need to protect the technologies (machineries and processes) used therein.[16] Similar

[13] Shirin Elahi, *Scenarios for the Future: How Might IP Regimes Evolve by 2025? What Global Legitimacy Might Such Regimes Have?* (European Patent Office 2007) 15.
[14] Penrose (n 3) 53.
[15] Manteaw (n 1) 114–15.
[16] George Sipa-Adjah Yankey, *International Patents and Technology Transfer to Less Developed Countries: The Case of Ghana and Nigeria* (Avebury 1987) 104–06.

patterns are found in Kenya, where the IP regime was introduced when it became a British colony in 1897. The extension of the British IP regime to Kenya was not meant to promote local creativity and innovation since it occurred when the levels of literacy and technological advancement among the native population were extremely low and local innovation virtually non-existent.[17]

It appears evident that, in general, behind the introduction of the IP systems in European states during medieval times there was a strong desire to attend to national interests and the need to operationalize national policies that could promote technological and industrial progress and the welfare of their own citizens.[18] This is in direct opposition to what was witnessed when the European powers were extending their IP systems to the newly conquered territories outside Europe. In that case, there was only the desire to extend privileges and benefits to European citizens who were spread across the world, as the history of the introduction of patent law in Ghana clearly illustrates. However, in those territories there were also indigenous people to whom those privileges were not extended, nor could they easily benefit from the IP system. Undeniably, that would have been possible only if the specific needs of the local people at that time were considered. Such an approach could not have been expected from the colonial masters as this was not among their priorities. It is indeed telling that German legislation expressly prevented the natives from holding IP rights in its colonies in Africa.[19]

An assessment of the modalities of the extension of IP systems to the African countries further illustrates how IP represented an alien imposition in the African construct.

3. Modalities of Introduction of the Intellectual Property Systems in Africa

The introduction of the IP system on the African continent was rather prescriptive. According to Kongolo, the colonial masters introduced the system in Africa in three distinct ways, namely 'the extension system'; the system of enactment of IP laws on behalf of the colonies; and the mixed approach, which incorporated elements of 'the extension system' and also allowed the colony the adopt its own specific IP legislation.

---

[17] Patricia Kameri-Mbote, 'Intellectual Property Protection in Africa: Assessment of the Status of Law, Research and Policy Analysis on Intellectual Property Rights in Kenya' (International Environmental Law Research Centre (IELRC) 2005) Working Paper No 2, 5 <http://erepository.uonbi.ac.ke/handle/11295/41242> accessed 14 May 2017.

[18] Pasquale Joseph Federico, 'Origin and Early History of Patents' (1929) 11 JPatOffSoc'y 292, 292–93.

[19] Alexander Peukert, 'The Colonial Legacy of the International Copyright System' in Ute Röchenthaler and Mamadou Diawara (eds), *Copyright Africa: How Intellectual Property, Media and Markets Transform Immaterial Goods* (Sean Kingston Publishing 2012) 40.

The first regime, the extension system, allowed colonial powers to merely extend the application of their own IP laws to the colonies through ordinance, decree, or order.[20] An example of this mechanism was the (Colonial) Gold Coast Patent Ordinance no 1 of 1899.[21] Patent Ordinance no 1 of 1899 was subsequently replaced by the Patents Registration Ordinance of 1925 and the Patents Registration (Amendment) Decree of 1972.[22] According to these legal instruments, patents could only be registered in the UK and re-registered in Ghana.[23] It is evident that the IP legal and institutional framework was not designed to build either local IP expertise or a 'culture' of IP in Ghana. To be sure, the colonial system was never conceived as giving the local people any possibility of devising inventions and innovations worthy of protection. In the unlikely event that such a desire arose, section 4 of Cap 179 of the Patents Registration Ordinance, 1925 was so ill-conceived that it required the residents of the colonies to send their applications to Europe before they could be granted and only thereafter could their rights be extended to their own state in Africa through the system of re-registration of the patent previously granted in the UK.[24] It is hence a reasonable inference that local innovation was not encouraged by the colonial IP system as devised.

The same pattern was followed in Nigeria: Patents Ordinance (No 30) of 1916, amended in 1925, which became the Registration of United Kingdom Patents Ordinance (No 6), provided for the registration in Nigeria of patents which had been granted in the UK.[25] As in Ghana, even a local inventor had first to apply to the Patents Office in the UK for the grant of the patent, before he could have his patent registered in Nigeria.[26] In the case of industrial designs, the United Kingdom Designs (Protection) Ordinance (No 36) of 1936 that provided for the protection in Nigeria of the designs registered in the UK did not even require any re-registration in Nigeria: the industrial design once registered in the UK was automatically valid

---

[20] Tshimanga Kongolo, 'Historical Developments of Industrial Property Laws in Africa' (2013) 5 WIPO J 106.

[21] Manteaw (n 1) 114–15.

[22] George M Sikoyo, Elvin Nyukuri, and Judi W Wakhungu, 'Intellectual Property Protection in Africa: Status of Laws, Research and Policy Analysis in Ghana, Kenya, Nigeria, South Africa and Uganda' (African Centre for Technology Studies 2016) 16 <www.jstor.org/stable/resrep00103.1> accessed 29 October 2023.

[23] Section 4 of Cap 179 of the Ordinance provided: 'Any person being the grantee of a patent in the United Kingdom, or any person deriving his right from such grantee by assignment, transmission or other operation of law, may apply within three years from the date of issue of the patent, to have such patent registered in the Gold Coast.'

[24] See text of the provision (n 23).

[25] The British first introduced the patent system in the former colony of Lagos and Southern Nigeria in 1900 by the Patents Ordinance No 17 of 1900 and the Patents Proclamation Ordinance No 27 of 1900, respectively. The Patents Proclamation Ordinance No 12 of 1902 introduced similar legislation in Northern Nigeria. Following the amalgamation of Southern and Northern Nigeria in 1914 the separate legislation for the different regions was repealed and substituted by the Patents Ordinance No 30 of 1916, which was amended in 1925 to become the Registration of United Kingdom Patents Ordinance No 6 of 1925. See Sikoyo, Nyukuri, and Wakhungu (n 22) 19.

[26] Adejoke Oyewunmi, *Nigeria Law of Intellectual Property* (Unilag Press & Bookshop 2015) 146.

in Nigeria. It did not matter if that design was irrelevant to the country or could even be contrary to the public interest or local customs and culture. Clearly, industrial design extended to Nigeria in this way had no regard for the circumstances prevailing in the country and was not intended to provide any contribution to the local economy.[27]

In the Republic of Zambia (then Northern Rhodesia), section 2 of the English Law (Extent of Application) Act Chapter 11 of the Laws of Zambia provided that, subject to the provisions of the Constitution of Zambia and any other written law, common law, doctrine of equity, and statutes that were in force in England on 17 August 1911 (the date of commencement of the Northern Rhodesia Order Council 1911), any statute of later date in force in England applied thereto by any Act or otherwise shall be in force in the Republic.[28] Intellectual property laws as adopted and changed in England had automatic and full validity in Zambia without question, irrespective of whether those laws could be beneficial to the country.

The Kenya Patent Registration Ordinance (1933) provided the legal grounds for the protection of patents in Kenya during the colonial era. In terms of section 54 of the Patents Registration Act, Cap 508, in order to obtain patent protection, an applicant was required to present a certified copy of letters patent from the UK Patents Office. That allowed re-registration of the patent in Kenya without further examination.[29] Application for re-registration had to be made within three years from the date of the grant and the patent would remain in force only for as long as the patent remained in force in the UK.

The French African colonies were governed by the following French laws, as amended: Patent Law of 5 July 1844, Trademark Law of 28 June 1857, and Industrial Design Law of 14 July 1909. The application of these laws was extended to the colonies by ordinances or orders. The Patent Law was extended to the colonies by an order issued by the President of the Council of Ministers on 21 October 1848 in conformity with article 51 of the Patent Law of 1844, which explicitly states that ordinances could be taken to regulate the application of the law in the colonies, with amendments as necessary. Since 19 April 1951, the central body for issuing patents, trademarks, and industrial designs was the French National Institute of Industrial Property (INPI) and it granted the rights without substantive examination.

The first regime of administration of industrial property rights was also applied by Portugal. The Patent Law of Portugal clearly stated that patents, utility models, and industrial designs confer the exclusive right to exploit the invention throughout the Portuguese territory, including all colonies.[30] However, in the

---

[27] ibid 208.
[28] George M Kanja, *Intellectual Property Law* (The University of Zambia 2006) 216.
[29] James Otieno Odek, 'The Kenya Patent Law: Promoting Local Inventiveness or Protecting Foreign Patentees?' (1994) 38 JAL 79, 79.
[30] Kongolo (n 20) 108.

Lusophony there was a specific regime with regard to trademarks as the protection resulting from the registration of a trade mark was limited to the metropolitan and insular territory and the protection could be extended to the colonies only after completion of the required formalities.[31]

The second regime allowed colonial powers to enact, on behalf of the colonies, IP laws that were applicable only in the colonies. The spirit of these laws in most cases did not depart much from the main laws adopted by the colonial power for the metropolitan territory. They were customized laws aimed at responding to specific interests of the colonial power in the colonies in general or in particular areas such as agriculture, biodiversity, and commerce. Under this scenario, the application of these adopted IP laws was limited to the territory of the defined colonies.[32] This second scenario corresponds to the bizarre Belgian method whereby the colony would adopt industrial property laws, similar to the metropolitan laws, on behalf of the colony that would be applicable only in that territory. For example, King Leopold II of Belgium enacted a Decree on Patents on 29 October 1886 and a Decree on Industrial Designs on 24 April 1922. For the implementation of those decrees, the Administrator General of the Ministry of Foreign Affairs issued an Instruction (order) Relating to Patents and Trademarks on 23 May 1889. According to article 1, 'a patent application in Congo must be lodged with the Governor General, who shall forward it to the Ministry of Foreign Affairs'. On receipt of the application by the Ministry of Foreign Affairs, a patent was prepared and signed by the Administrator General on behalf of the King. It would then immediately be sent to the Congo to be remitted to the patentee. The Governor General, through the Ministry of Foreign Affairs, had to be notified of any assignment or transfer of title made in the Congo.[33]

The third regime, according to Kongolo, was a mixed approach that would incorporate elements of the 'extension system' and could also contain exceptions allowing the colony to adopt its own IP legislation. Therefore, the colony could regulate patents and industrial designs by extending the main laws on the subject matter to their colonies while simultaneously adopting distinct trade mark legislation, enacted specifically for the colony.[34]

4. The Administration Structures

The dependence of the African IP systems was also evident in the way that the administration of the IP system was structured. It is well established that during the

---

[31] ibid.
[32] ibid 107.
[33] ibid.
[34] ibid.

colonial period, the IP systems in Africa were built around, and to serve, the IP offices of the colonial masters such as France, Portugal, Spain, or the UK. The IP offices in Africa were designed to serve as mailboxes for the local applications to be channelled to the European capitals and to re-register the rights already granted in the metropolitan European states. It was claimed that, since the African IP offices lacked technical expertise and patent applications were very few, the complex work of examining and granting patents could only be done in Europe, thus removing the burden and the need to establish examination offices or even develop a local capacity to administer patent applications in Africa. This clearly impeded the development of skills for examination and granting of patents in African colonies and kept the monopoly of that knowledge within the European patent offices. The effect of such practice is still felt across the entire continent of Africa up to the present day, and it is very difficult to have any confidence that this gloomy scenario will change soon.

These findings portray a series of misfortunes that affected the process of the introduction of an IP system in Africa: if the African governing systems were unable to develop a legal framework to promote and protect technological progress, the colonial system did not seek to improve the situation. On the contrary, the colonial system supressed any chance of the local innovation system thriving by promoting western-style innovation protected by an IP system that was biased against indigenous knowledge. All these factors blocked the process of innovation and technological progress on the continent.

C. The Intellectual Property System in the 'Non-colonized' African States

1. Context

The colonial system influenced the path of progress, including technological development, on the African continent. It is not the purpose of this study to dwell on the debate about whether such influence brought any positive contribution to technology progress or should only be seen as the element that disrupted the local processes of innovation. However, it is worth considering that there were a few African countries that were not subjected to the phenomenon of colonialism, or at least not in the harsh form witnessed in the majority of colonies. The two states that fall under that category are Ethiopia and Liberia.

Regarding the few exceptions, the question that arises is whether those countries were able to nurture their own technological progress and to what extent the utilization of local knowledge contributed to furthering their own objectives of development.

2. The Case of Liberia

As in all other African states, Liberia was inhabited by a population belonging to local tribes with a long history and traditions. The contribution of the indigenous populations to the evolution of humankind and technological progress in its infant stages falls under the general submission made in Chapter 2. However, what makes Liberia unique are the events that occurred at the beginning of the nineteenth century, soon after the abolition of slavery with the 'Back to Africa' movement sponsored by the American Colonization Society, which promoted the return of the black populations from America to Africa.[35]

In 1822, the American Colonization Society (formed in 1816 by eminent persons from several states of the Union (North America)) founded a settlement on the west coast of Africa that they called Monrovia.[36] The American Colonization Society obtained land from, among others, indigenous leaders of the Dei, the Gola, the Condo, the Grebo, and the Kru tribes through a one-off payment of cash (Spanish dollars) and goods (such as brass-barrelled pistols, cocked hats, and fishhooks); in some cases, pieces of land were even appropriated without any treaty.[37] The objectives of the Society were to establish a civilized Christian state in West Africa by repatriating liberated slaves from the United States of America (US) and to provide them with the means of settlement, subsistence, and defence.[38] Subsequently, other colonization societies were formed in some states of the Union, such as Maryland, New York, Pennsylvania, and Mississippi, and these societies founded other settlements near Monrovia and attracted more freed negroes from America.[39] Later, the various settlements formed the Commonwealth of Liberia.

This new entity on the African continent assumed a *sui generis* status in the international arena in that, although Liberia is considered to be the only colony ever established by the US, it was never recognized as such and no attempts were made to assume control or responsibility over the territory. Liberia simply remained a private individual enterprise occupying a large territory over which no sovereign government claimed or exercised jurisdiction.[40] Subsequently, Liberia declared itself an independent state and adopted a constitution modelled after that of the US on 26 July 1847.[41]

---

[35] According to Kazanjian, these were white nationalists who were also abolitionists, with the help both of proslavery forces who wanted a way to deport free blacks and rebellious slaves, and of some free blacks who had given up hope of living a free life in the United States, see David Kazanjian, 'The Speculative Freedom of Colonial Liberia' (2011) 63 AmQ 863, 864.
[36] JH Mower, 'The Republic of Liberia' (1947) 32 J NegroHist 265, 265.
[37] It must be emphasized, however, that colonists were discouraged from invoking spirits to acquire land, Kazanjian (n 35) 865.
[38] Mower (n 36) 266.
[39] By the end of the nineteenth century more than 16,000 colonists settled in Liberia.
[40] Unknown, 'Liberia' (1909) 3 AJIL 958, 960.
[41] ibid.

Due to the excellent climate and soils, all settlements that later formed Liberia were established on the firm foundation of an agrarian economy, and Liberia's seal is also clear evidence of reliance on agriculture.[42] However, lack of capital, skills, equipment, and draft animals, and inadequate use of agricultural techniques hindered development of agriculture by colonists in Liberia. To mitigate some of these challenges, the Liberian government requested support from the US in order to introduce oxen and horses, but with limited success. The introduction of some crops with export value such as tobacco, coffee, sugar, and cotton also yielded limited success.[43]

Drawing specifically on technology transfer and progress it is worrying that the depiction of the seal mimics rudimentary equipment developed by North Americans in the area of agriculture, but which was not adapted to Liberian conditions and not improved further. Apparently, the focus on an agrarian economy also limited the horizons of technology progress in Liberia.

With regard to the legal system, it is noteworthy that colonists were dispatched with a constitution, including a bill of rights, modelled after the North American legal instruments.[44] In the circumstances, Liberia did not develop any remarkable legal framework by itself and although its IP legislation dates back to 1864 through the Liberian Patent Act of 23 December 1864, similarities with the IP legislation that existed in the US demonstrates that it was imported from that country.

Liberia struggled to provide for its own needs and on many occasions pleaded for support from the US, which was reluctant to provide it. France and Great Britain pledged friendship agreements and protectorate status and even attempted to annex Liberia on several occasions.[45] It was in these circumstances that Liberia participated at the Berne Conference in 1886 and became one of the original signatories to the Berne Convention for the Protection of Literary and Artistic Works of 1886. In light of this background it is hard to believe that Liberia represented the indigenous people of that country. Evidence shows that the government of Liberia participating at the Berne Conference was made up of the black colonists inspired by the legislation in force in the country, which was fully influenced by the American culture and legislation on IP. With regard to technology, the very same dichotomy is observed whereby the indigenous Liberian people followed exactly the same patterns witnessed in other states in Africa and especially the 'conservatism' that blocked technological progress while the black colonists originating

---

[42] Samuel W Laughon, 'Administration Problems in Maryland in Liberia: 1836–1851' (1941) 26 JNH 325, 328; Kazanjian (n 35) 864. It is to be noted that the national seal of Liberia depicts a dove on the wing with an open scroll in its claws, a view of the ocean with a ship under sail, the sun just emerging from the waters, a palm tree, and at its base a plough and spade. Beneath the emblems, the words REPUBLIC OF LIBERIA appear, and above the emblems, the national motto, THE LOVE OF LIBERTY BROUGHT US HERE.
[43] Laughon (n 42) 350.
[44] ibid 326.
[45] Mower (n 36) 268–70.

from America were trying to surmount the challenges of introducing rudimentary technology learnt in the American farming industry during slavery.

As a result, although the history of Liberia is unique in that it remained one of the few territories in Africa over which no imperial power formally claimed sovereignty, it did not find its own path to development. The settlement of black colonists freed from slavery in the US added a further complication in the identity of the nation. Therefore, the early independence of Liberia did not provide any competitive advantage to the people of Liberia, and there is no outstanding contribution in the affirmation of local knowledge in Africa, its mechanisms of protection, or any home-grown technological progress that Liberia and Africa can be proud of.

3. The Case of Ethiopia

Some of the oldest fossils of human-like creatures dated as six to seven million years old were found in Ethiopia. Pre-humans such as the *Ardipithecus kadabba* (5.8–5.6 million years old) and *Ardipithecus ramidus* (4.4 million years old) were also found in Ethiopia.[46] More importantly, the well-known *Australopithecus afarensis* (from the Afar triangle) which came to be known as 'Lucy', estimated to have lived 3.2 million years ago, was also found in northern Ethiopia. Evidence of the birth of modern humans on the African continent is also based on the discovery of fossils of *Homo sapiens* 165,000–195,000 years old in Ethiopia's Lower Omo and Middle Awash valleys.[47] Similarly, evidence of the first expression of the human capacity to develop technology through stone artefacts (oldowan tools) was found in Ethiopia.[48] Indeed, the earliest lithic technology was found in records located at Kada-Gona, Kada-Hadar, and Omo-Shungura in Ethiopia with an estimated age of 2.6 million years.[49]

It is relevant at this stage to note that Ethiopia is also one of the oldest states in the world. Thus, the foundations of Ethiopia as a state can be traced back to Axum, which flourished from 5 BCE to 16 CE.[50] The foundations of the medieval Ethiopian state can be found in the Kingdoms of Lalibella (1137–1270), Shoa (1300–1600), and Gondar (1632–1885). Prominent leaders in the medieval history of Ethiopia are identified as Amde Tsion (1314–44), Negus Yeshaque (1414–29), Zere Yacob (1434–68), and Serse Dingil (1563–97).[51]

---

[46] Yves Coppens, 'Hominid Evolution and the Emergence of the Genus Homo' (2013) 9 ScriVaria 2.
[47] John Shea, 'Refuting a Myth About Human Origins' (2011) 99 AmeSci 135.
[48] Coppens (n 46) 17.
[49] Eudald Carbonell and others, 'Eurasian Gates: The Earliest Human Dispersals' (2008) 64 JAR 195, 209.
[50] Daniel Kendie, 'The Causes of the Failure of the Present Regime in Ethiopia' [2003] IJES 179.
[51] ibid 180.

For the purposes of this book, it is worth assessing the case of Ethiopia since it managed to sustain an unbroken chain of historical civilization free of foreign 'corruption'.[52] Indeed, Ethiopia is considered unique in Africa due to its uninterrupted freedom, usually explained through the isolation paradigm.[53]

Kendie asserts that medieval Ethiopians knew, worked, and used iron, and that Ethiopians were skilled in the arts and architecture. Ethiopian soldiers were armed in the same fashion as early European medieval soldiers, and churchmen were known by their piety and learning.[54]

While refuting any idea of technological developments in medieval Ethiopia, Aregay observes that Ethiopia had traded with Arabia, India, and the Mediterranean world since the beginning of the fourteenth century to sell them slaves, ivory, musk, wax, and coffee in exchange for swords, helmets, silk, brocades, cushions, and carpets.[55] This therefore rejects the isolation paradigm as the main factor behind the failure of Ethiopia to develop or adopt better technology. Aregay instead finds that the fundamental reason behind such failure was the inability or reluctance of the ruling classes to develop traditions and institutions for the secure ownership and transmission of property and offices for the benefit of the general population.[56]

A number of emperors who ruled Ethiopia, recognizing the lack of the skills necessary to develop the technology, made an effort to introduce and adopt foreign technology. In particular, Emperor Yeshaq, who ruled from 1413 to 1430, encouraged craftsmen from Egypt and as far afield as Spain to migrate and settle in Ethiopia, and on many occasions they were not allowed to leave the country.[57] Notwithstanding these policies, few good craftsmen flocked to Ethiopia or remained there, although there is an account of Circassian Mameluke, a master armourer who is said to have produced mail and weapons for the emperor and his army. However, it is intriguing that a century later, Emperor Lebnä Dengel (ruling from 1508 to 1540) imported all military equipment from Egypt, thus demonstrating that Ethiopia did not make any effort to learn from the earlier technology, improve on it, and further develop its own technology. In the chronicles of Alvarez it emerges that the Emperor had a great interest in guns and gunpowder, swords, and European clothes.[58] However, the lack of innovativeness on the part of the ruler never inspired him to request the various foreigners to reproduce the technology and to develop those industries within Ethiopia.

---

[52] Teshale Tibebu, 'Ethiopia: The "Anomaly" and "Paradox" of Africa' (1996) 26 JBS 414, 414.
[53] Arnold Toynbee, *A Study of History*, vol 2 (G Cumberlege, Oxford University Press 1946) 180.
[54] Kendie (n 50) 180.
[55] Aregay (n 4) 127.
[56] ibid.
[57] ibid 128.
[58] Francisco Alvares, *The Prester John of the Indies* (Published for the Hakluyt Society at the University Press 1961).

In this regard, Calestous Juma emphasizes the role of leaders as risk-takers on behalf of the public.[59] Unwise leaders may instead forego support for the introduction of new technologies due to a possible lack of support by the masses, even if the probability of those technologies solving societal challenges is high. The ability of leaders to allow the introduction of new technological solutions while maintaining continuity, social order, and stability may seal the fate of the society. To that end, it is necessary to have what Juma designates as entrepreneurial leaders:

> capable of using the available knowledge to assess the situation, take informed executive action in a timely manner, and continue to monitor technological advances and their impacts.[60]

Instead, what Ethiopia witnessed was a limited vision of the ruling class which prevented Ethiopian society from developing traditions and institutions which would have protected the property of merchants and craftsmen, hence fostering individual creativity.[61]

Subsequent periods of Ethiopian history did not improve the situation, and in some cases there was even regression. The modern Ethiopian state found its solid foundations in the nineteenth century. The building blocks of modern Ethiopia were laid by the Emperors Tewodors II (1855–68), Yohannes IV (1872–89), Menelik II (1889–1913), and Haile Selassie I (1930–74).[62] With respect to this period, Lewis terms Ethiopia 'mysterious, isolated, diminished but unconquered'.[63] It was indeed during the rule of Emperor Menelik II that Ethiopia defeated the Italians at the battle of Adwa on 1 March 1896. The Adwa victory was considered 'the greatest single disaster in European colonial history' and substantially enhanced Ethiopia's respectability among western powers; Ethiopia was subsequently recognized as a sovereign state.[64] Notwithstanding the victory of Adwa, Ethiopia remained behind with regard to economic development.[65]

Italy successfully invaded Ethiopia in October 1935 and occupied the territory up until 1941 when, under the leadership of Emperor Haile Selassie, it was liberated. Selassie is credited with having ended the long period of isolation by joining the United Nations in 1945, the Non-Aligned Movement in 1961, and for making Addis Ababa the headquarters of the United Nations Economic Commission for Africa (1958) and the Organization of African Unity (1963).[66]

---

[59] Calestous Juma, *Innovation and Its Enemies: Why People Resist New Technologies* (Oxford University Press 2016) 7.
[60] ibid 282–83, 286.
[61] Aregay (n 4) 133–43.
[62] Kendie (n 50) 180.
[63] David L Lewis, *The Race to Fashoda: European Colonialism and African Resistance in the Scramble for Africa* (1st edn, Weidenfeld Nicolson 1987) 416.
[64] Tibebu (n 52) 418.
[65] ibid 420.
[66] Kendie (n 50) 184.

Unfortunately, some of these achievements, such as those in the areas of education and health, did not reach the majority of Ethiopians who were living in the rural areas: the ratio of primary school pupils to total population was only 1.68 per cent and in healthcare, the ratio was 0.34 hospital beds per 1,000 people.[67] Ethiopia therefore remained backward with regard to technological progress compared to the most developed countries of the western world.

Haile Selassie was overthrown in 1974 by Mengistu Haile Mariam ,who ruled from 1977 to 1991. His military regime adopted a 'socialist' orientation that established a monopolistic party structure. As a result, all old patterns were deliberately destroyed. Traditional values and institutions were replaced by a new set of institutions, laws, and standards that followed the soviet communist model. Consequently, the major means of production, including rural land and urban houses, were seized and became state owned. Furthermore, tenant and landlord relationships were abolished, and private ownership of the means of production disappeared.[68]

With regard to IP, up until 1995, Ethiopia did not have any specific legislation to deal with patents, utility models, and industrial designs; did not join any international legal instrument related to IP up until 2002; and there were no specialized institutions dealing with the matter.[69] The first IP law was enacted on 10 May 1995, with accession to the World Intellectual Property Organization only occurring in 1997. Only in 2003 was an autonomous IP office established.[70] For the first time, Ethiopia recognized the importance of IP in promoting technological progress and set as the objective of the new legislation the creation of favourable conditions in order to encourage local inventions and related activities, thereby building up national technological capability and encouraging the transfer and adaptation of foreign technology. Notwithstanding these timid developments, Ethiopia is still characterized by a phobia of international treaties, as evidenced by the fact that up until now Ethiopia has not yet become a signatory of the most basic international treaties such as the Paris Convention for the Protection of Industrial Property of 1883, the Berne Convention for the Protection of Literary and Artistic Works of 1886, and the Patent Cooperation Treaty of 1970.

Ethiopia will remain the pride of Africa and the symbol of defiance of western domination, and it will always be inspirational in terms of African identity. However, it is disappointing that this unique state which enjoyed 'freedom from

---

[67] ibid.
[68] ibid 185.
[69] Joseph M Wekundah, 'A Study on Intellectual Property Environment in Eight Countries: Swaziland, Lesotho, Mozambique, Malawi, Tanzania, Uganda, Kenya and Ethiopia' (African Technology Policy Studies Network (ATPS) 2012) African Technology Policy Studies Network Working Paper Series No 66, 33–40 <www.africaportal.org/publications/a-study-on-intellectual-property-environment-in-eight-countries-swaziland-lesotho-mozambique-malawi-tanzania-uganda-kenya-and-ethiopia/> accessed 12 January 2019.
[70] ibid 33.

foreign corruption' did not capitalize on its strength to develop a model that could showcase the capacity of the African continent to encourage the use of its indigenous knowledge to promote development and technological progress. It is also disappointing that the ruling class did not take advantage of local and foreign skills to promote adoption and improvement of foreign technologies in order to foster innovation. Most importantly, it is clear from the discussion that the ruling class did not create conducive conditions and further failed to set up policies and a legal framework and systems that would enable innovation to thrive in Ethiopia.

The analysis of the examples of the two African 'non-colonized' states was revealing with regard to the real reasons behind the failure of technological progress on the African continent. Although colonialism had significant influence in halting progress on the continent, this study finds that there are other reasons embedded in African culture, the governmental systems, and the rulers' approach to technology and science that are seriously hindering development. Those factors also explained why, even after fifty years of independence of other African states, technological progress and innovation are not yet forthcoming. If any state in Africa is one day to overcome the challenges of technological progress and innovation it will have to engage in serious introspection into the root causes of its sociological systems, culture, government, and policies. As already elaborated, that process is possible and has taken place elsewhere in the world, specifically in China. China was able to overcome deep-rooted neo-Confucian and Taoist traditions, coupled with decades of a debilitating communist regime exerting power, to become a flag-bearer of the patent system today. That change was possible due to the government's deliberate policies that placed IP centre stage of its efforts to promote technological progress, innovation, and a knowledge-based economy. This is also possible in Africa, as will be discussed in Chapters 5 and 7.

D. Post-Independence Implementation of Intellectual Property Systems in Africa

1. The Role of Intellectual Property in Africa's Post-Independence

The main concern for African states soon after independence was the rapid industrialization of their territories. To achieve that objective, they followed a model that was thriving in Latin American countries, based on an import substitution strategy.[71] African states were particularly interested in the domestic production of goods that were mostly imported at that time in order to attend to the basic needs of their people. This included mainly non-durable consumer goods, some

---

[71] Ana Paula F Mendes, Mário A Bertella, and Rudolph FAP Teixeira, 'Industrialization in Sub-Saharan Africa and Import Substitution Policy' (2014) 34 BrazilJPolEcon 120, 121.

intermediate goods, and very few capital goods.[72] In the long run, the desire was to shift from the production of consumer goods to more sophisticated products. Although local production required some level of technological development, not all goods that were manufactured locally under the drive of import substitution required innovation. Nevertheless, this chapter looks at the approach of African states in regard to industrialization and innovation with a view to determining whether they used IP as a tool to foster industrial development, technological progress, and innovation soon after their independence. Furthermore, this chapter assesses the extent to which IP is integrated into developmental strategies and policies, such as industrial, science, technology, and innovation (STI) policies. As a way of breaking the ground for future initiatives, this chapter analyses some best practices in mainstreaming IP policies into national development programmes through the examples of Mozambique, Rwanda, and South Africa.

2. Africa's Post-Independence Industrialization Policies and Intellectual Property

The developmental policies adopted by African states after independence were largely concerned with rapid industrialization of their territories with a view to meeting the needs of their people. By the time the majority of the African states attained their independence in the 1950s and 1960s they were highly dependent on European metropoles with regard to the export market, technology, management and entrepreneurship capabilities, capital, and the economic system in general.[73] The economic system was indeed heavily linked to the former colonial masters and unintegrated into the neighbouring states. This appears to be one of the intricate legacies of colonialism. For example, British colonies in West Africa, namely Gold Coast (now Ghana), Nigeria, Gambia, and Sierra Leone, were integrated among themselves with a common currency, services, and an airline company. On the other hand, the thirteen French colonies were also integrated as a francophone family through two federations, namely the French West Africa and French Equatorial African Federations.[74] As a result, the two blocs of states were linked to their colonial masters, but less connected to the other African states outside their language or colonial enclave.

In addition, during the colonial era, manufacturing activities were discouraged by the authorities.[75] Initial stages of manufacturing that happened in Africa around

---

[72] ibid 128; Overseas Development Institute, *Industrialization in Sub-Saharan Africa* (1986) 3.

[73] William Steel and Jonathan Evans, 'Industrialization in Sub-Saharan Africa—Strategies and Performance' (World Bank 1984) 13.

[74] Afua Boatemaa Yakohene, 'Overview of Ghana and Regional Integration: Past, Present and Future' in Friedrich Ebert Stiftung (ed), *Ghana: In Search of Regional Integration Agenda* (Friedrich Ebert Stiftung 2009) 4.

[75] Steel and Evans (n 73) 7.

1920 focused exclusively on the processing of agricultural raw materials for export, such as palm oil from Cameroon, Nigeria, and Ghana; lime juice from Ghana; sawmilling from Ivory Coast; and sisal, cotton, coffee, and tea from Tanzania.[76] Only very limited industrialization, having as its purpose to serve European elites living in the colonies, was introduced, such as baking, food processing, beverages, tobacco, textiles, shoes, and wooden products. Both international agents and local colonial governments viewed industrialization policies in the colonies as a threat to the established monopoly of production of manufactured goods which was reserved for the metropolis and the supply of cheap raw materials to the same market from the colonies.[77] Of relevance is the fact that industrialization was allowed within the colonial framework and the industries were all owned and managed by Europeans, and to some extent by Asian and Middle Eastern nationals, while Africans were barred by law from owning companies or because of the lack of the necessary education, skills, and capital.[78] As a result, qualified human resources were scarce and limited to some professions and the public service.

In view of these circumstances, at the time of independence the prevailing scenario in African states was characterized by a primary exporting economy focused on minerals and crops, lacking an entrepreneurial class, with no capital, relying on unqualified people with limited knowledge of technology.[79] As a result, newly independent governments vigorously reacted to this scenario and vowed to alter their economic systems and explore new forms of production. They endeavoured to achieve this by emulating the successful import substitution model followed by Latin America states to address their challenges.[80] In carrying out import substitution, African states started with non-durable consumer goods, such as foodstuffs, drinks, tobacco, and textiles; intermediate goods such as rubber, chemicals, oil, metal, and non-metallic mineral products; and capital goods such as electrical machines and transportation equipment.[81] The long-term strategy of these states was to subsequently switch the export structure from basic items to more expensive goods.[82] From the second half of the 1960s onwards, African governments took full control of the industrial development system and policies with the goal of assuring the realization of their vision.[83] In Ghana, for example, a Seven-Year Development Plan designed to develop domestic industry, infrastructure, and social welfare was promoted by the government of Kwame Nkrumah.[84] To achieve his vision,

---

[76] ibid.
[77] Mendes, Bertella, and Teixeira (n 71) 121.
[78] Steel and Evans (n 73) 8.
[79] Mendes, Bertella, and Teixeira (n 71) 125.
[80] ibid 126.
[81] ibid 128; Overseas Development Institute (n 72) 3.
[82] Mendes, Bertella, and Teixeira (n 71) 126.
[83] ibid 129–30.
[84] Government of Ghana, 'Seven-Year Development Plan: 1963/64–1969/70' <https://ndpc.gov.gh/media/Ghana_7_Year_Development_Plan_1963-4_1969-70_1964.pdf> accessed 9 November 2018.

Nkrumah's government nationalized the major industries as well as the agricultural sector.[85] Between 1966 and 1971, President Houari Boumédienne of Algeria promoted the nationalization of 90 per cent of the country's industries, including oil, gas, and industrial production, with a view to using the profits to develop state-owned industrial enterprises.[86] In Kenya, between 1963 and 1978, President Jomo Kenyatta promoted the programme of state-managed capitalism with import substitution industrialization that protected the domestic market for consumer goods, and imported essential capital goods to promote the development of industry.[87] In Ethiopia, Emperor Haile Selassie undertook a strategic industrial development from 1957 (through three major five-year development plans until 1974) and used import substitution industrialization to promote agricultural commercialization and development of the textile and beverage industries.[88]

It is fair to acknowledge that the state-led development programmes rolled out in Africa in the 1960s and 1970s produced significant results, as evidenced by the growth in gross domestic product (GDP) and comprehensive social welfare programmes.[89] Some of the policies adopted in foreign trade policy included preferential treatment for capital and intermediate goods as well as certain basic inputs; the imposition of administrative import control through quotas, licences, and tariffs; and price control and prohibition of imports of goods similar to those produced domestically.[90] With regard to investment policy, governments opted for large investments in manufacturing, development of basic infrastructure for the industry, and establishment of public institutions devoted to industrial development.[91]

As a result of these policies throughout the continent, industrialization certainly expanded and factories mushroomed for the large-scale production of textiles, paint, plastics, light drinks, and beer; construction materials (such as ceramics, faucets, pipes, floor tiles, and roof tiles); pharmaceuticals; fertilizers and agro-industrial products; exploration of minerals, such as iron; and the production of oil- and petroleum-based products.[92] At the initial stage, data appeared very positive in favour of these policies, as evidenced by the exponential growth of the number of state-owned companies in Ghana, Ivory Coast, Mali, Mauritania, Senegal,

---

[85] 'Capitalism—Africa—Independence, State-Led Development, and Import-Substitution Industrialization—African, Economic, Women, and Program—JRank Articles' <https://science.jrank.org/pages/8525/Capitalism-Africa-Independence-State-Led-Development-Import-Substitution-Industrialization.html> accessed 9 August 2017.
[86] ibid.
[87] Nicola Swainson, 'State and Economy in Post-Colonial Kenya, 1963–1978' (1978) 12 CJAS/RCEA 357, 365–67.
[88] Mulatu Wubneh, 'State Control and Manufacturing Labor Productivity in Ethiopia' (1990) 24 JDA 311, 310–13; Dejene Aredo, 'Developmental Aid and Agricultural Development Policies in Ethiopia 1957–1987' (1992) 17(3) AfrDev 209–37.
[89] Wubneh (n 88) 310–13; Aredo (n 88) 3.
[90] Mendes, Bertella, and Teixeira (n 71) 129.
[91] ibid.
[92] ibid.

Tanzania, Uganda, and Zambia. Significantly, the share of state-owned companies in the GDP of Sub-Saharan African states rose to 17.5 per cent.[93] Nigeria is highlighted as the country that achieved the highest added value due to manufacturing, while Kenya, Ivory Coast, Cameroon, and Zimbabwe showcased significant results in promoting import-substituting industries.[94] However, this positive picture soon faded as the growth in size was not followed by better production. Indeed, it was revealed that only Nigeria, Kenya, and Zambia became industrialized and among them only Nigeria and Kenya maintained a good performance.[95]

Among many obstacles to the successful implementation of these policies were lack of human capital and little knowledge of technology.[96] For example, it has been highlighted that the agricultural sector did not benefit from public investment in technological innovation and farmers continued growing crops using outdated practices and technology, rendering them unable to support industrial growth.[97] Further, shortage of personnel and lack of human capital adversely affected the business of industries. Most relevant is the fact that governments did not focus on issues related to technological innovation; hence, although the introduction of new technologies to improve and increase production and distribution of goods and services was found relevant, no specific technological policy and planning were set out in this area.[98] The case of Nigeria is clear evidence of this failure. Nigeria adopted its First National Development Plan (1962–68) and introduced import substitution industrialization to encourage industrial development and lessen dependence on foreign trade, thereby conserving foreign exchange by producing local products that were previously imported. The main objective of the import substitution industrialization strategy was to stimulate the start-up and growth of industries, as well as to enhance indigenous participation. However, it led to high technological dependence on foreign know-how, to the extent that the domestic factor endowments of the country were grossly neglected.[99]

In assessing this context, it was clearly revealed that the desire to achieve rapid industrialization prompted African states to opt for procurement of technology in developed states and to establish capital-intensive industry instead of exploiting

---

[93] The number of state-owned companies in Tanzania, for example, rose from eighty in 1967 to 400 in 1981. In Kenya, they increased threefold, from twenty—at the time of independence in 1963—to sixty in 1969. In Ghana, there were no state-owned companies at its independence in 1959, yet there were approximately 100 in the early 1960s. In other countries, such as Zambia, Senegal, Ivory Coast, Mali, Mauritania, and Uganda, there was a marked increase in the number of those companies, ibid.
[94] Overseas Development Institute (n 72) 4.
[95] Mendes, Bertella, and Teixeira (n 71) 131.
[96] ibid.
[97] ibid.
[98] ibid 133.
[99] Louis N Chete and others, 'Industrial Policy in Nigeria: Opportunities and Challenges in a Resource-Rich Country' in Carol Newman and others (eds), *Manufacturing Transformation: Comparative Studies of Industrial Development in Africa and Emerging Asia* (Oxford University Press 2016); Udo Ekpo, 'Nigeria Industrial Policies and Industrial Sector Performance: Analytical Exploration' (2014) 3 IOSRJEF Journal 1, 2–4.

the available labour force at the national level. It was also clear that states needed skilled labour to manage the capital-intensive technology which was not available on the continent and was supplied by developed states. Therefore, it is submitted that although modest results of industrialization were initially achieved, ownership of the process and transfer of skills and technology did not occur. This was due to the fact that governments did not focus on creating local capacity to sustain the industrialization process but were merely concerned with the production of non-durable consumer goods that could substitute imports and satisfy local needs.

It is interesting to note is that development policies adopted during the sixties, seventies, and eighties also did not mention the role of IP in promoting development. Consequently, the real impact of IP in economic and technological development in the first three decades of independence of African states was almost nil. Furthermore, IP institutions on the continent were almost non-existent or were not active and the IP legal framework was still what had been inherited from the colonial masters, which did not consider the real conditions of African states and did not respond to the pressing need for development of the continent.

Contrary to the African scenario where technology importation was an end in itself, as described, the situation in Europe has been always different. It must be recalled that in Europe, technology importation was devised to promote industrial development, innovation, and the development of IP systems. The IP systems introduced since the medieval period in Europe were specifically devised to attract foreign skilled artisans, to facilitate importation of technology, or to induce local people to invent devices for the common good of society. In England, as early as 1326 a policy of encouraging the importation of new arts from abroad was enacted, and in 1337 a statute was introduced establishing that all cloth workers of other states would be given special franchises and privileges if they settled in England and practised and taught their arts to the British citizens.[100] This is unequivocal evidence that already during medieval times, governments in Europe and especially in England had identified the patent system as an important tool to promote the establishment of new industries and acquisition of technology.

A quick perusal of the situation in the developed world reveals that the trend continued throughout unabated. Specifically, states that were technologically backward could rise on technologies developed in other states and could craft their IP systems in a way that would allow their citizens and firms to catch up in terms of the latest technological advancements. In some circumstances that included denying IP rights to foreigners in order to benefit their own citizens, as evidenced by the case of the US, which denied copyright protection to foreigners to facilitate the growth of its publishing industry.[101] This policy was reversed and the protection

---

[100] Federico (n 18) 293.
[101] Nirmalya Syam and Viviana Muñoz Tellez, *Innovation and Global Intellectual Property Regulatory Regimes: The Tension Between Protection and Access in Africa* (South Centre 2016) 15.

was granted to foreigners only after America believed that it had achieved its objectives at the end of the nineteenth century.[102] Similarly, in 1869 the Netherlands abolished patent protection to avoid Philips infringing Edison's patents and to enable it to produce light bulbs.[103] The development of the chemical and textile industries in Switzerland in the nineteenth century is also attributed to a strategic decision to refuse patents to foreigners who were far ahead in terms of skills in that sector.[104] The rapid industrial development of Japan is also credited to a patent system that was deliberately crafted to facilitate importation, indigenization, and diffusion of foreign technology.[105] To that end, Japanese IP legislation could force foreign patent applicants to disclose their technological knowledge even before the grant of a patent, or they could grant a short grace period that made it harder for foreign firms to file patent applications.[106]

3. Intellectual Property in the Context of the Current Industrialization and Innovation Policies

The value of technology and innovation in promoting development and improving productive capacities and people's lives cannot be overemphasized.[107] Accordingly, policymakers, scientists, the private sector, and civil society have recognized that science, technology, and innovation are the drivers of African development in the knowledge era. Consequently, efforts have been made in recent years to pair innovation and industrial policies. Considering the fact that innovation usually goes hand in hand with IP, it is to be expected that industrial and innovation policies would also incorporate IP as the main driver. It is therefore relevant at this stage to assess whether IP policies were used to promote industrial development and foster innovation in Africa. The cases of Kenya and Nigeria have been selected in view of their apparent success in developing their industrial sectors in recent years. However, these two states were also relevant because they showcased typical instances of African states that are not taking advantage of the IP system and have not adopted a national IP policy.

---

[102] ibid.
[103] ibid.
[104] ibid.
[105] ibid 16.
[106] ibid.
[107] Engwa Azeh Godwill, 'Science and Technology in Africa: The Key Elements and Measures for Sustainable Development' (2014) 14 GJSFR: G BIO-TECH & GEN 16, 21; Stephen Young and Ping Lan, 'Technology Transfer to China through Foreign Direct Investment' (1997) 31 RegStud 669, 670; Keith Maskus, 'Encouraging International Technology Transfer' [2004] IPRSD 1, 7; Anja Breitwieser and Neil Foster, 'Intellectual Property Rights, Innovation and Technology Transfer: A Survey' (The Vienna Institute for International Economic Studies 2012) 88, 47 <https://wiiw.ac.at/p-2646.html> accessed 24 April 2023; Surendra J Patel, 'The Technological Dependence of Developing Countries' (1974) 12 J ModAfrStud 1, 4.

a) Kenya

Kenya adopted the 'National Industrialization Policy Framework for Kenya, 2012–2030' in November 2012.[108] The policy aspires to transform Kenya into a middle-income rapidly industrializing state and focuses on value addition for both primary and high-value goods. It also aims to develop linkages between industrial sub-sectors and other productive sectors to drive the industrialization process and provide strategic direction for the sector's growth and development.[109] The policy defines three industrial sectors as priority areas for development: labour-intensive industries; medium to high technology sectors; and advanced manufacturing technologies.[110] Although the policy acknowledges that an effective IP rights system is an incentive to innovation, the policy measures suggested lean more towards the enforcement of the rights as an end in itself. These include the enactment of the East African Anti-Counterfeit Bill, restructuring and strengthening of the Intellectual Property Tribunal, strengthening of the Anti-Counterfeit Agency and other institutions involved in the fight against counterfeit goods, and, in general, strengthening institutional capacity for the administration of IP rights.[111] The policy therefore focuses more on protection of existing rights than promoting innovation. One will recall that developed states emphasize the relaxation of IP rights so as to promote innovation notwithstanding that little protection is granted by these states, as clearly indicated earlier in this chapter.

Kenya's national policy ultimately recommends fast-tracking the enactment of a binding national IP policy.[112] Several drafts have been drawn up but not yet adopted.[113] However, one can expect that such a policy will favour protection more than innovation. Kenya has also adopted innovation policies. Kenya's first innovation policy was launched in 2006. In 2009, a comprehensive policy on Science, Technology, Innovation Policy and Strategy (STIPS) was developed for purposes of mainstreaming the application of STI and innovation in all sectors and processes of the economy.[114] Furthermore, Kenya has enacted the Science, Technology and

---

[108] Ministry of Industrialization, 'National Industrialization Policy Framework for Kenya 2012–2030: Transforming Kenya into a Globally Competitive Regional Industrial Hub' <https://repository.kippra.or.ke/handle/123456789/1037> accessed 18 June 2017.

[109] Policy objective 4.7 on the standards, quality infrastructure and intellectual property rights, ibid 25–32.

[110] The three priority areas include: labour-intensive sectors such as agro-processing, textiles and clothing, leather and leather goods; medium- to high-technology sectors, namely iron and steel, machine tools and spares, agro-machinery and farm implements, and pharmaceuticals; and, as a third area, advanced manufacturing technologies, which include biotechnology and nanotechnology, Ministry of Industrialization (n 108).

[111] ibid 34.

[112] ibid.

[113] David Opijah, 'Towards a National Intellectual Property Policy for Kenya' <www.jdsupra.com/profile/david_opijah_docs/> accessed 14 October 2017.

[114] Bitange Ndemo, 'Effective Innovation Policies for Development: The Case of Kenya' in INSEAD and WIPO (eds), *The Global Innovation Index 2015: Effective Innovation Policies for Development* (WIPO 2015).

Innovation Act No 28 of 2013.[115] The Act establishes the National Commission for Science, Technology and Innovation in section 3.1. Section 6.2 establishes that the Commission shall have powers to apply for the grant or revocation of patents; institute such action in respect of the patent as it may deem appropriate for national security; and acquire from any person the right in, or to, any scientific innovation, invention, or patent of strategic importance to the state. Under the same law, the Kenya National Innovation Agency has responsibility for increasing awareness of IP rights among innovators; developing national capacity and infrastructure to protect and exploit IP derived from research or financed by the Agency; and to facilitate applications for the granting or revocation of patents, as well as instituting legal action for infringement of any IP rights.[116]

From this elaboration, it appears evident that the innovation and industrial policies in Kenya neglect the important role that IP can play in fast-tracking innovation and industrial development and instead prioritize the enforcement segment.

b) Nigeria

Another state that is worth assessing is Nigeria. According to World Bank data, Nigeria hosts the largest economy in Africa. In 2023, its population was more than 223 million, with a GDP of over US$500 billion. However, half of the GDP is generated by primary sectors such as agriculture, while the oil and gas sector is the major driver of the economy, which in 2011 made up over 95 per cent of export earnings and about 85 per cent of government revenue.[117] To change this scenario the government adopted an agenda for economic transformation known as 'Nigeria Vision 20:2020'; a long-term plan for stimulating Nigeria's economic growth and launching the country onto a path of sustained and rapid socio-economic development.[118] The industrialization strategy incorporated in this vision aims to achieve greater global competitiveness in the production of manufactured goods by linking industrial activity with primary sector activity, domestic and foreign trade, and service activity. The vision also incorporates elements of relevance to STI. A national science and technology policy was formulated and launched in 1986 with the aim of placing science and technology at the centre of national development. The policy emphasized the importance of promoting the transfer of technology from foreign to local firms.[119] Pertinent here is the Nigeria Industrial Revolution Plan (NIRP) adopted in 2014, labelled as the first strategic, comprehensive, and integrated roadmap to industrialization, designed as a five-year plan to accelerate the development of industrial capacity in Nigeria. The plan aimed to

---

[115] Kenya Science, Technology and Innovation Act 2013.
[116] ibid 28.1.
[117] Chete and others (n 99) 99.
[118] 'Nigeria Vision 20:2020' <www.nigerianstat.gov.ng/pdfuploads/Abridged_Version_of_Nigeria%20Vision%202020.pdf> accessed 10 November 2018.
[119] Chete and others (n 99) 123.

increase manufacturing's contribution to GDP from 4 per cent to 6 per cent by 2015, and ambitiously, above 10 per cent by 2017.[120] Data available reveals that although there was substantial improvement, oil still dominates Nigerian GDP and the contribution of manufacturing remained below 10 per cent in 2018.[121]

The Nigeria Industrial Revolution Plan is based on the desire to drive a process of intense industrialization, based on sectors where Nigeria has a comparative advantage.[122] The NIRP only mentions IP with respect to 'support structures and enablers'. Although it alludes to IP under the section related to 'innovation', the remedy prescribed is related to the improvement of IP enforcement procedures to protect knowledge assets.[123] This is further reinforced by the prescription to control counterfeiting and fakes, which could harm consumers and negatively impair branded foods in the area of agribusiness.[124] Regarding manufacturing, IP is once again called upon to deal with the issue of counterfeiting.[125] The policy indicates that Nigeria has already established the broad framework for adequate protection of IP rights, but it is in the enforcement of these rights that the challenge lies, as evidenced by the rampant piracy and counterfeiting levels in Nigeria. Therefore, it is evident that IP is only viewed as tool to discourage copying and counterfeiting, and not as a powerful instrument to foster innovation and the much-desired industrial revolution.

The NIRP therefore missed the great opportunity to mainstream IP into the industrial revolution of Nigeria. The industrial revolution is disconnected from IP and has no role in value addition in the natural resources of Nigeria, and thus it is difficult to understand how the linking of industrial activity with primary sector activity will happen. This reality reinforces the thesis of this book that IP needs to be incorporated into the national development policies. Of concern in Nigeria is the fact that the development of an IP policy, although launched, is proceeding at a rather slow pace. It is strongly submitted in this book that an IP policy is indeed fundamental to building the vision of the state in relation to the use of IP for development and to chart the path for its concrete operationalization.[126]

---

[120] 'Nigerian Industrial Revolution Plan (NIRP)' <www.nipc.gov.ng/product/nigerian-industrial-revolution-plan-nirp/> accessed 10 November 2018.

[121] 'Nigeria's Economy: Services Drive GDP, but Oil Still Dominates Exports' (*Africa Check*, 2018) <http://africacheck.org/fact-checks/reports/nigerias-economy-services-drive-gdp-oil-still-dominates-exports> accessed 12 March 2020.

[122] Such as the agro allied sectors; metals and solid minerals related sectors; oil and gas related industries; as well as construction, light manufacturing, and services.

[123] 'Nigerian Industrial Revolution Plan' (n 120) 112.

[124] ibid 34.

[125] ibid.

[126] Femi Olubanwo, Oluwatoba Oguntuase, and Banwo Ighodalo, 'Strengthening Intellectual Property Rights and Protection in Nigeria—Trademark—Nigeria' (11 March 2019) <www.mondaq.com/nigeria/trademark/788714/strengthening-intellectual-property-rights-and-protection-in-nigeria> accessed 15 October 2019.

Similar scenarios to those illustrated in Kenya and Nigeria abound on the African continent. Accordingly, African states are urged to consider integrating IP into their national development policies and to avoid viewing the enforcement of IP rights as an end in itself. There may be light at the end of the tunnel and some hope that this may be corrected, judging by some initiatives to develop IP policies spearheaded by the World Intellectual Property Organization (WIPO) under its technical assistance programmes.

4. Some African Best Practices in Mainstreaming Intellectual Property Policies into National Development Programmes

a) Policymaking space for IP policy development

African states perform poorly on the generation and ownership of IP rights, as evidenced by statistics that have demonstrated clearly that Africa is not taking advantage of the IP system.[127] As a result, there are calls for Africa to make use of the policymaking space that still remains to adopt the most appropriate policies that can assist the continent in catching up in terms of technological progress and innovation.[128] WIPO provides technical assistance for the development of national IP policies in the developing and Least Developed Countries (LDCs). WIPO postulates that IP is an important component of national economic policies. However, it is observed further that governments, especially in developing and LDCs, are confronting challenges in adequately designing national IP systems that best serve their policy objectives.[129]

One of the forty-five recommendations of the Development Agenda of WIPO is to assist states to develop and improve national capacity by making national IP institutions more efficient and promoting a fair balance between IP protection and the public interest.[130] Pursuant to this, as a member-driven organization, and at the request of national governments, WIPO has been supporting the efforts of formulating IP policies. A significant aspect of this important intervention is the acknowledgement that notwithstanding the fact that the IP system has been considerably harmonized, there is no 'one size fits all' approach in designing IP policies.

Despite WIPO having received a wide range of requests from member states for assistance in the formulation and implementation of IP policies in the recent past—and the fact that it is eager to provide such assistance under the WIPO Development

---

[127] UNECA, 'Assessing Regional Integration in Africa VII: Innovation, Competitiveness and Regional Integration' 93–94 <https://repository.uneca.org/handle/10855/23013> accessed 8 October 2017.
[128] ibid 96.
[129] WIPO, 'Methodology for the Development of National Intellectual Property Strategies—Tool 1: The Process' <www.wipo.int/edocs/pubdocs/en/wipo_pub_958_1.pdf> accessed 8 April 2019.
[130] Recommendation 10 of 'The 45 Adopted Recommendations under the WIPO Development Agenda' <www.wipo.int/ip-development/en/agenda/recommendations.html> accessed 17 October 2017.

Agenda projects—it is frustrating to note that very few states have leveraged that opportunity. Indeed, only Botswana, Cape Verde, Malawi, Mozambique, Namibia, Rwanda, South Africa, Zambia, and Zimbabwe have adopted national IP policies and strategies.[131] Ethiopia, Kenya, and Uganda have been in the process of formulating their policies for many years without much success, and in all other states there is no evidence of such a process being envisaged.

The IP policies developed in Mozambique, Rwanda, and South Africa are further analysed so as to assess their possible impact on national development policies and priorities.

b) Rwanda's Intellectual Property Policy

One of the pioneer countries in Africa to adopt a national IP policy, with the support of WIPO, is Rwanda. Through this policy, Rwanda sought to articulate a vision; define policy objectives and specified strategies for ensuring the effective use of IP as a tool for development; and support the goals of Vision 2020 and objectives of various policies such as the STI policy, the education policy, and the industrial policy.[132] The Rwandan Intellectual Property Policy is remarkable in that it provides direction with regard to IP in relevant areas. For example, in utility models it indicates that this tool may provide the most important avenue for using IP to support development. Utility models are viewed as particularly relevant in achieving policy objective II—encouraging minor and incremental innovation and creativity. The policy expressly excludes pharmaceutical products from patentability in accordance with the WTO Decision providing for a transition period for LDCs until at least 2033.[133] The policy also affirms that consideration should be given to retaining patent examiners to enforce the requirements of enabling disclosure under IP law in key sectors such as agriculture, even though Rwanda does not intend to routinely examine all patent applications.[134] The policy also indicates that procurement should be strategically used in order to promote local brands, inventions, and

---

[131] Mozambique, 'Intellectual Property Strategy 2008–2018' <www.aripo.org/resources/mozambique-intellectual-property-strategy-2008-2018/> accessed 16 October 2017; Rwanda, 'Revised Policy on Intellectual Property in Rwanda' <https://www.newaripo.online/storage/resources-member-state-policies/1674822545_phpHJDmGy.pdf> accessed 29 October 2023; Zimbabwe, 'The Zimbabwe National Intellectual Property Policy and Implementation Strategy 2018–2022' <https://www.newaripo.online/storage/resources-member-state-policies/1674822688_phpRksdBI.pdf> accessed 29 October 2023; Zambia, 'Revised National Intellectual Property Policy 2020' < https://www.pacra.org.zm/wp-content/uploads/2021/08/RevisedNationalIntellectualPropertyPolicy.pdf> accessed 29 October 2023; South Africa, 'Intellectual Property Policy of South Africa—Phase I (2018)' <www.gov.za/documents/intellectual-property-policy-south-africa-%E2%80%93-phase-i-2018-13-aug-2018-0000> accessed 19 March 2019.

[132] WIPO, 'Integrating Intellectual Property into Innovation Policy Formulation in Rwanda' (2015) <www.wipo.int/publications/en/details.jsp?id=3943> accessed 16 October 2017.

[133] WTO, 'Extension of the Transition Period Under Article 66.1 of the TRIPS Agreement for Least Developed Country Members for Certain Obligations with Respect to Pharmaceutical Products' (adopted 6 November 2015) Doc IP/C/73 <www.wto.org/english/tratop_e/trips_e/art66_1_e.htm> accessed 16 November 2018.

[134] Rwanda, 'Intellectual Property Policy' (n 131) 18.

creative works.[135] In general, the policy indicates that all of these provisions will have to be implemented in an appropriate and balanced manner to ensure that they encourage innovative and creative activities and investment while also facilitating access to technology and essential goods and services.

Although not much detail is provided, the Rwandan Intellectual Property Policy may constitute a starting point for the development of specific instruments and incentives with a view to realizing the goals defined in it. However, the policy could have gone further to highlight how IP can fast-track innovation. For instance, the policy indicates that patent examiners should be retained in order to enforce the requirements of enabling disclosure under IP law.[136] Nevertheless, it was not clear from the policy point of view how the technological information that will be disclosed will be identified, collected, and utilized to foster innovation in agriculture. One positive aspect of the policy is that it emphasizes that achieving its objectives will require strategic implementation of associated policies and legislation and defines investment policies and government procurement as priorities. However, it was not immediately evident what the content of those policies will be in order to operationalize those statements. These are some of the issues that the remaining policy space must address in future.

c) Mozambique's Intellectual Property Strategy
Another country that has leveraged the opportunity provided by WIPO to assist in the formulation of national IP policies is Mozambique.[137] Mozambique's Intellectual Property Strategy takes into account the most important national, regional, and international instruments guiding the state's development, such as Agenda 2025, the Expanded Program for the Reduction of Absolute Poverty (PARPA), the government's five-year plan, the Millennium Development Goals, the New Partnership for Africa's Development (NEPAD), and the various policies and strategies of the relevant sectors in the IP rights field, particularly the Policy on Science and Technology; the Strategy for Science, Technology and Innovation; the Industrial Policy and Strategy; the Rural Development Strategy; the Policy on Traditional Medicine; the Cultural Policy; and the Strategic Plan on Education and Culture.[138]

[135] ibid.
[136] ibid.
[137] Mozambique, 'Intellectual Property Strategy 2008–2018' (n 131) 131.
[138] Mozambique, 'Agenda 2024: Nation's Vision and Strategies' <www.mef.gov.mz/index.php/documentos/83-agenda-2025/file?force-download=1> accessed 10 November 2018; Government of Mozambique, 'Program for the Reduction of Absolute Poverty—PARPA 2001–2005' <www.mef.gov.mz/index.php/documentos/7-parpa-i/file?force-download=1> accessed 10 November 2018; Government of Mozambique, 'Program for the Reduction of Absolute Poverty 2006–2009' <www.mef.gov.mz/index.php/documentos/7-parpa-i/file?force-download=1> accessed 10 November 2018; 'Programa Quinquenal Do Governo 2010–2014' <www.portaldogoverno.gov.mz/por/content/downl oad/1959/15690/version/1/file/Plano+Quinquenal+do+Governo+2010-14.pdf> accessed 15 August 2018; 'Programa Quinquenal Do Governo 2015–2019' (BR I Série N° 29 de 14 de Abril de 2015) <www.mef.gov.mz/index.php/todas-publicacoes/instrumentos-de-gestao-economica-e-social/programa-qui

Mozambique's Intellectual Property Strategy views IP as an instrument to stimulate and protect creativity and innovation to promote the state's economic, scientific, technological, and cultural development. It defines priority areas and illustrates in detail how IP has the potential to fast-track value addition in those areas. It has also set up an institutional framework to facilitate implementation of the strategy, with specific reference being made to establishing a funding system. However, the IP strategy lapsed in 2018 without being implemented in a coordinated manner and its impact in the development of the IP system in Mozambique seems to be marginal.

d) South Africa's Intellectual Property Policy

The formulation of an IP policy in South Africa has been slow and the first Draft Intellectual Property Policy released in September 2013 was fiercely resisted due to a lack of consultation with the relevant stakeholders.[139] The latest version of the Draft Intellectual Property Policy was finally released in August 2017 after extensive consultation and was finally adopted in 2018.[140] The adopted policy describes IP as an important policy instrument in promoting innovation, technology transfer, research and development (R&D), creative expression, consumer protection, industrial development, and, more broadly, economic growth.[141] The policy draws on the understanding that knowledge, innovation, and technology are increasingly becoming the drivers of progress, growth, and wealth. Based on that assumption, the policy highlights that one of the core elements of the structural conditions necessary to enable South Africa to make the transition towards a knowledge economy is an IP policy.[142] The new Intellectual Property Policy is designated as a 'phase I' policy and has targeted the priorities of the state by focusing on measures related to IP and public health.[143] Significant in this regard is the statement in the policy that:

> It should be recalled that IP is an instrument of industrial policy that is tailored by state organs to accomplish development objectives. [...] As nations adjust their industrial policy, including in relation to social policy, so too do they adjust the rights and obligations of IP holders. In line with the South African Constitution, a balanced approach will be taken in the development of the IP Policy.[144]

nquenal-do-governo-pqg/doismilequinze-doismiledezanove/797-balanco-do-pqg-2015-2019/file?force-download=1)> accessed 18 September 2019.

[139] Linda Daniels, 'South Africa Approves New IP Policy, with Guidance from UN Agencies' *Intellectual Property Watch* (Geneva, 6 June 2017).
[140] South Africa, 'Intellectual Property Policy of South Africa—Phase I (2018)' (n 131) 131.
[141] ibid 132.
[142] ibid 9.
[143] ibid 4, 14.
[144] ibid 10.

This statement was found to be laudable as it seems to be targeting the current IP debates in the South African context, namely its impact on public health. Secondly, it is significant that the policy is deliberately linked to industrial policy and will be tailored to the local context in order to craft a balanced system. As such, the South African Intellectual Property Policy seems to be a decisive move in the desired direction since it has defined a narrow and clear aspect of IP that the state intends to tackle: public health and access to medicines. As highlighted, the Intellectual Property Policy's overall objective is to stimulate genuine innovation and to achieve that, it has defined the introduction of substantive search and examination procedures for patents as one of the key reforms to be undertaken.[145] The Policy highlights that this will benefit patent-holders by granting them rights that were thoroughly assessed, and, at the same time, will benefit the public at large by ensuring that market exclusivity is only granted when appropriate.

e) South Africa substantive examination priority areas
However, considering the limited capacity of South Africa to conduct substantive examination in all sectors, the state has decided to prioritize examination in the area of pharmaceutical patents. This is in line with the 2014 'Policy Guide on Alternatives in Patent Search and Examination' developed by WIPO that illustrates the various options available to states for the search and examination of patent applications and guides them in selecting the best option suitable for their state.[146] The Policy Guide states that 'addressing capacity constraints can be done by dividing the substantive examination into specific technological fields that the state will prioritise. Applications relating to other fields of technology may be subject to formality examination only or outsourcing substantive examination either within or outside' the state.[147] The South African Intellectual Property Policy elucidates that, as capacity is developed, the fields which will be subject to full substantive patent examination will also be expanded after consultation with stakeholders.[148] Given the scenario provided in Subsection D.4 d) on 'South Africa's Intellectual Property Policy', it was submitted that the policy represented a good example of a formulation process that is participatory, takes into account the local context, and prescribes solutions to the concrete issues that are relevant for the state using the most appropriate mechanisms. It was therefore argued that this example should be emulated elsewhere on the continent.

---

[145] ibid 5.
[146] WIPO, 'Policy Guide on Alternatives in Patent Search and Examination' 8 <www.wipo.int/edocs/pubdocs/en/wipo_pub_guide_patentsearch.pdf> accessed 18 October 2017.
[147] ibid.
[148] South Africa, 'Intellectual Property Policy of South Africa—Phase I (2018)' (n 131) 5.

E. Conclusion

The extension of the IP systems developed and implemented in Europe to Africa did not support the development of indigenous knowledge but instead promoted its marginalization. However, although the dominant narratives about Africa blame colonization for the current status of technological development of African states, stagnation in the levels of development and technological uptake were also found in non-colonized African states, as evidenced by the cases of Ethiopia and Liberia. Therefore, a much deep introspection is required to identify the endogenous factors that are impeding the technological development of Africa apart from the legacy of colonization.

The dawn of decolonization did improve the situation significantly, in view of the fact that IP remained on the margins of the innovation ecosystems and in the technological and economic development policies crafted after the independence of African countries.

The examples of Rwanda and South Africa illustrate the ideal direction that the process of formulation of IP policies should follow in Africa. The policies, as exemplified in Rwanda and South Africa, should have clear objectives with regard to technological development and innovation and should be mainstreamed with other relevant policies. In addition, IP policies should provide concrete directives on how to employ IP to foster innovation as shown in the case of the policies in these two states.

This chapter also highlights the three important elements that constitute an advancement on the minimum steps for establishing an appropriate IP policy. South Africa represents an example of best practice that could be emulated because the policy is realistic in acknowledging the difficulties in defining directives and implementing a comprehensive framework in all areas. Therefore, it chose to take a phased approach whereby attention is first placed on the topical issue that is currently of greatest concern in the IP arena in South Africa, namely public health. The second commendable aspect of the South African Intellectual Property Policy is that it is informed by evidence. The policy sets out the context with sufficient information about the situation in South Africa and at the regional and international levels, as well as including an analysis of possible constraints and opportunities. Thus, the policy choices that were taken seem to be based on concrete scientific information. Formulating an evidence-based IP policy is crucial to achieving the desired results. However, it is worth acknowledging that some African states may not be able to afford to undertake the exercise of collecting evidence for policy formulation as has been done in South Africa due to their limited resources. An

CONCLUSION 83

alternative would be to rely on the technical assistance provided by relevant developmental agencies to fund or prepare the expert research.[149] Lastly, the wide consultation and transparency in the process of formulation of the Draft Intellectual Property Policy in South Africa is likely to be a fundamental ingredient that will facilitate ownership of the document by all sectors and enable its successful implementation.[150]

Based on the observations made, it is suggested that the process of formulating and implementing IP policies should emulate the positive experiences of the three states analysed. An inadequate process of formulating an IP policy, lack of scientific evidence, insufficient consultation, lack of coordination between the relevant institutions, and failure to define priority areas to be targeted by the policy may threaten the success of the process of implementing IP policies. It is advisable that the important policy space provided by the ongoing process of formulating IP policies in Africa should observe these vital elements if they are to achieve positive outcomes.

---

[149] Caroline Ncube, 'The Development of Intellectual Property Policies in Africa: Some Key Considerations and a Research Agenda' (2013) 1 J IntellectPropRights 3.

[150] A Consultative Framework was approved by the South African government in July 2016 allowing contributions from all stakeholders. The Draft Intellectual Property Policy was finalized and launched in August 2017, and stakeholders were granted sixty days to provide further inputs.

PART II
THE USE OF INTELLECTUAL PROPERTY TO FOSTERING DEVELOPMENT IN AFRICA

4
The Challenges of Promoting Innovation through Technology Transfer into Africa

A. Introduction

A systematic analysis of the endeavours of African states to use intellectual property (IP) to drive innovation and industrialization, as well as the challenges thereof, was conducted in Chapter 3. In this chapter, the focus is primarily on the discussion of the philosophical justifications of IP protection and how this can be reinterpreted to foster innovation and development. Based on some of the positions, it can be understood why the international community prioritized the establishment of a legal regime that enables technology transfer for the benefit of disadvantaged states, including those in Africa.

According to the rights-based approach, development should be based on rights rather than on benevolence or charity simply provided through technical assistance.[1] Therefore, the duty of charity that was initially formulated by Locke must be converted into a legal obligation for it to be effective in assisting developing states to reframe their claims into legitimate entitlements.[2] For the rights-based approach, mere charity is no longer enough; instead, it is imperative to shift the perspective from viewing development as a need, to the stage where it is considered a fully-fledged right.[3] Translating this philosophical construct into a practical tool to promote technology transfer and innovation for the benefit of developing states requires a clear international legal framework. However, as early as the 1960s it was observed that an international legal framework to govern the issue of technology transfer was sorely lacking.[4] In view of that, some attempts were made to establish such an international regime, but without much success. This chapter illustrates the current international regime of technology transfer and innovation and explores ways of improving it. The chapter provides an overview of the attempts to establish a legal regime, namely through the United Nations Conference

---
[1] Paul Gready, 'Rights-Based Approaches to Development: What Is the Value-Added?' (2008) 18 DevPrac 735, 742–44.
[2] ibid; Jaakko Kuosmanen, 'Repackaging Human Rights: On the Justification and the Function of the Right to Development' (2015) 11 JGE 303, 303–20.
[3] Aili Mari Tripp, 'Development and the New Rights-Based Approaches in Africa' (2009) 36 ROAPE 279, 279; Brigitte I Hamm, 'A Human Rights Approach to Development' (2001) 23 HRQ 1005, 1026.
[4] Susan K Sell, *Power and Ideas: North-South Politics of Intellectual Property and Antitrust* (State University of New York Press 1997) 199.

on Trade and Development (UNCTAD) Draft International Code of Conduct for the Transfer of Technology of 1976.[5] It also illustrates how the vacuum left by the Draft Code of Conduct paved the way for several international legal instruments to mushroom, leading to the fragmentation of the international technology transfer regime. It further describes the proposal by Barton and Maskus to adopt a Multilateral Agreement on Access to Basic Science and Technology (ABST) and the attempt to partially include it in the WIPO Development Agenda.[6]

A historical approach further revealed that there was an evolution in the efforts to address issues related to transfer of technology. A breakthrough was achieved, in particular, by the Kyoto Protocol to the United Nations Framework Convention on Climate Change, which includes some provisions on international technology transfer. Following this legal instrument, around thirty multilateral agreements followed suit to address technology transfer issues, mainly in the environmental protection domain. These developments led some scholars to suggest that the technology transfer provisions of the Kyoto Protocol and other international treaties could be assembled to compose a multilateral, legally binding agreement—the International Transfer of Technology Agreement (ITT).

Notwithstanding these efforts, a comparative analysis of those regimes still shows gaps. This is because, in the first place, more than half of those instruments focus on environmentally sound technologies and hence have limited scope.[7] Apparently, there are no compelling reasons to justify the exclusion of all other areas of technology in the new regime. Secondly, the adopted instruments are soft law and implementation by the states that acceded to those international instruments depends on their willingness to do so. This chapter leverages these important milestones on the long journey aiming at establishing an international regime for technology transfer and proposes an omnibus and binding instrument. This seems to be a unique formula to finally provide a definitive solution to the lack of a unitary framework to govern the international technology transfer regime that the world has been seeking in the last fifty years. Such an instrument would be revolutionary and would provide for the first time, after so many attempts, a harmonized regime to facilitate transfer of technology in the world and especially from developed countries to developing and Least Developed Countries (LDCs). Details of that regime will be further elaborated in Chapter 6.

The chapter is structured as follows: an introductory section provides an overview of the philosophical debates that justify transfer of technology from producing countries to those in dire need. This is followed by the analysis of the UNCTAD Draft International Code of Conduct for the Transfer of Technology, including its

---

[5] UNCTAD Secretariat, 'Draft International Code of Conduct on the Transfer of Technology' <https://digitallibrary.un.org/record/86199> accessed 13 May 2018.

[6] Keith Maskus, 'Encouraging International Technology Transfer' [2004] IPRSD 1, 38.

[7] Zhong Fa Ma, 'The Effectiveness of Kyoto Protocol and the Legal Institution for International Technology Transfer' (2012) 37 JTT 75, 77.

failure which prompted the fragmentation of the legal regime. A number of legal instruments that attempted to regulate technology transfers separately are also described. The chapter further reviews the systematic approach to tackle fragmentation provided under Section D, 'Proposed Reforms on the International Framework for Regulating Technology Transfer' which discusses ABST, including the abortive attempt to implement it through the WIPO Development Agenda and the ITT Agreement. The chapter concludes by proposing the adoption of an omnibus and binding instrument on technology transfer.

B. Philosophical Debates on the Reinterpretation of Locke's Labour Theory to Facilitate Transfer of Technology and Foster Innovation

The central undertaking in this book involves an analysis of the extent to which the current IP legal framework prioritizes the establishment and consolidation of a strong IP protection system, while simultaneously perpetuating the exclusion of African states. Underlying this critical analysis is the determination of whether the existing framework can be transformed so as to advance Africa's quest for access to technology through the technology transfer mechanism. The issue is whether, if correctly implemented or improved, especially taking into account ideals of justice as developed by Locke and the development theory, the Agreement on Trade-Related Aspects of Intellectual Property Rights (TRIPS Agreement), which is often viewed as the stumbling block, could create an enabling environment for innovation in Africa.

Finding the answer to this question necessitates engaging with the philosophical debate surrounding the matter. It is first relevant to analyse the critique by Gordon of the prevailing interpretation of the Labour Theory as proposed by Locke.[8] Gordon affirms that Locke's Labour Theory of property and allied approaches have been used so frequently as justification for creators' ownership rights that Locke's *Two Treatises* has erroneously been credited with having developed an explicit defence of a robust IP system.[9] However, Gordon argues that the same theory could be used to limit those rights. He achieves this by summarizing the Lockean theory into four propositions:

a) All persons have a duty not to harm others;
b) All persons have the freedom to dispose of their efforts as they see fit, and all persons have the freedom to use the common property developed by others;

---

[8] Wendy J Gordon, 'A Property Right in Self-Expression: Equality and Individualism in the Natural Law of Intellectual Property' (1993) 102 YaleLJ 1533, 1540.
[9] ibid.

c) Each person has a duty to let others share in her resources (other than her body) in times of great need, so long as the sharer's own survival is not imperilled by such charity, and each has a duty to share any of their nonbodily resources, which would otherwise spoil or go to waste;
d) All persons have a duty not to interfere with the resources others have appropriated or produced by labouring for the common good.[10]

From these propositions Gordon derives what he calls 'moral claims and entitlements' or 'fundamental human entitlements' of which he highlights three: the right to be free from harm, to have a share of others' plenty in times of need, and the liberty to make use of the common property developed by others.[11]

Gordon submits that although these entitlements allow the labourer to possess his products undisturbed, natural law imposes on him an obligation to share his plenty with those in extreme need. As a result of this duty, in combination with the 'waste' limitation that requires that one not allow one's property to perish uselessly, the labourer's freedom to use the product may be limited.[12] Fundamental in this analysis is the Lockean proviso:

> Labour being the unquestionable Property of the Labourer, no Man but he can have a right to what that is once joined to, at least where there is enough and as good left in common for others.[13]

Therefore, according to Gordon, the creators have rights to their own endeavours but only if such grant of property does not deprive others of the opportunity to also create or to benefit from existing knowledge.[14] It is worth recalling that Gordon clarified that this limitation is also applicable to intangibles.[15] Gordon then provides the conceptual clarification of 'waste' as follows:

---

[10] ibid 1542.
[11] ibid 1543.
[12] ibid 1551.
[13] ibid 1562.
[14] Gordon's conclusion is exciting:
> In sum, if there is only one culture (and whether technological or literary culture is at issue, the point is the same), a person who wishes to contribute to it usually is required to use the tools of that culture. Giving first creators ownership over any aspect of the culture, even if that aspect is newly created, may make a later creator less well off than he or she would have been without the new creation. Intellectual products, once they are made public in an interdependent world, change that world. To deal with those changes, users may have need of a freedom inconsistent with first creators' property rights. If they are forbidden to use the creation that was the agent of the change, all they will have to work from will be the now devalued common. The proviso eliminates this danger. It guarantees an equality between earlier and later creators. The proviso would thus ensure later comers a right to the broad freedom of expression, interpretation, and reaction which earlier creators had, a right which cannot be outweighed by other sorts of benefits (ibid 1570).

[15] Similarly, Hettinger says says: 'When owners of intellectual property charge fees for the use of their expressions or inventions, or conceal their business techniques from others, certain beneficial uses of

a) There is irrevocably unmet demand;
b) The goods to satisfy that demand already exist; and
c) Property claims prevent satisfaction of those demands.[16]

Gordon's views appear to be very appealing. However, it is worth highlighting that Gordon's assessment addresses the demand for tangible goods and its regime of ownership, hence the analysis was not generalized to include intangibles. Therefore, the framework developed by Gordon does not establish a general rule in society with regard to how knowledge, or specifically how technology, can be shared. Furthermore, it does not specify how the limitation of rights can be achieved and how individual owners can be forced to share their endeavours. Accordingly, the analysis has expanded to cover technology as a specific form of IP right. The focus of this chapter is on technology as the targeted endeavour and specifically the obligation to transfer technology by drawing the connection between Locke's proviso and the relevant provisions of the TRIPS Agreement that sets out that obligation.

Gordon's reinterpretation of Locke was further expanded by Hull.[17] Hull's thesis attempts to resurrect the spoilage proviso. Paraphrasing Locke, he emphasizes that the initial common goods are 'given to men for the support and comfort of their being'. In addition, God has given people 'reason to make use of it to the best advantage of life and convenience'.[18] As a result, Hull identifies two moral duties imposed on those who possess goods or rights. The first is a duty of charity, which imposes the obligation to provide for those who are in need and cannot provide for themselves.[19] The second duty forbids those who possess rights or goods to waste what they have created or produced.[20] Hull derives from Locke's work the mechanisms to improve people's life rather than a justification of stronger IP rights. He reaches that conclusion by linking his theory to theology and substantiating that Lockean property rights are adequately blended with concerns of benevolence towards others.[21] In view of that, Hull affirms categorically that Locke's theory can safely be used to limit property claims and justify claims of justice in the IP system. Hull does not reject the assertion that IP claims find ground in Locke's theory, but he

---

these intellectual products are prevented. This is clearly wasteful, since everyone could use and benefit from intellectual objects concurrently', see Edwin C Hettinger, 'Justifying Intellectual Property' (1989) 18 PhilosPublicAff 31, 31.

[16] Gordon Hull, 'Clearing the Rubbish: Locke, the Waste Proviso, and the Moral Justification of Intellectual Property' (2009) 23 PAQ 67, 81.
[17] ibid.
[18] ibid 68.
[19] ibid 80; Robert P Merges, *Justifying Intellectual Property* (Harvard University Press 2011). 'Locke says that people in desperate need have a claim on the assets held by legitimate owners. For Locke, the destitute have title to the goods they need to survive, even when those goods are otherwise legitimately held by others, either through valid original appropriation or a subsequent transfer from an original acquisition.'
[20] Hull (n 16) 79; Gordon (n 8) 1543.
[21] Hull (n 16) 86–87.

refuses to accept the current trend of considering that this theory can only be used to justify strong IP regimes. The author calls for more research to understand how Locke would have applied property regimes specifically in the context of intangible assets so as to derive benefits for all humanity.[22] This book responded to the call to understand how the analysis, which was crafted taking into account specific products of intellect, could be expanded to cover the full spectrum of knowledge creation. The concept of 'waste' was applied to technology to emphasize the obligation to share it before it is lost or underutilized. In this regard, the book demonstrated, paraphrasing Locke, that there is 'enough as good' in terms of technology in the developed world and that there is indeed an unmet demand for technology in developing countries, but property claims prevent satisfaction of that demand.

Cumulatively, Locke's duty of charity is combined with the rights-based approach to develop an obligation on developed countries to transfer technology to developing countries that need this technology but cannot develop it in isolation.

Reinterpretation of Locke's provisos is further strengthened by applying the right to development and more specifically its derivative rights-based approach in order to turn Locke's duty of charity into a binding obligation. For example, Andreassen considers that the right to development can be used by former colonies as a potential moral and legal justification to demand reparation and compensation for past injustices perpetrated by the former colonial powers.[23] The right to development can also be used in the quest for raising welfare standards of the impoverished people in those former colonies.[24] Andreassen also believes that the right to development can be enhanced in order to play a constructive role in the planning and implementation of development policies and practices. The author views this right as a programmatic right, or a goal right, whose main purpose is to permeate the designing (and appraisal) of national and international policies and programmes. It appears as if a solid platform has been established for African countries to base their philosophical and legal justification to be fairly integrated into the global innovation process by way of drawing upon the developmental approach. In pursuit of establishing this solid foundation, reliance was placed on the position of Uvin, who submits that the right to development grants both legal and ethical authority to developing countries to claim redistribution of resources.[25] Uvin's conclusions confer full legitimacy on developing countries to claim a review of the current prejudicial international arrangements to prompt better access to technology. Based on that assumption, developing countries can raise their claims in order to redress the current imbalance in respect of technology issues.

---

[22] ibid 87.
[23] Bård-Anders Andreassen, 'On the Normative Core of the Right to Development' (1997) 24 ForumDevStud 179, 182.
[24] ibid.
[25] Peter Uvin, 'From the Right to Development to the Rights-Based Approach: How "Human Rights" Entered Development' (2007) 17 DevPra 597, 598.

Cornwall and Nyamu-Musembi consider that the Declaration on the Right to Development introduced a radical change in the understanding of rights, and especially its duty-bearers, by placing emphasis on the global dimension.[26] They submit that the Declaration sets a collective obligation on all states to create a just and equitable international environment for the realization of the right to development.[27] The authors also encourage the global actors, such as the donor community, intergovernmental organizations, and international non-governmental organizations, to create monitoring and accountability procedures that will contribute to the enjoyment of human rights in each country and for the realization of universal human rights.[28] Cornwall and Nyamu-Musembi believe that relying on a set of internationally accepted legal documents is a more powerful and legitimate approach to development.[29]

The findings of these authors are very relevant to the debate on the promotion of innovation but are too general in nature. For this reason, this volume fine-tunes those findings and applies them specifically to the issue of promoting innovation in developing countries with a specific focus on African states.

Gready endorses Cornwall and Nyamu-Musembi's view by suggesting that a rights-based approach can reframe development into an entitlement secured through a political and legal contract not only with the state but also with other key actors and stakeholders.[30] The rights-based approach redefines the development work as being based on rights rather than on benevolence or charity channelled through technical assistance.[31] The rights-based approach as viewed by Gready can therefore contribute to development and can operationalize the right to development and other human rights. Gready's submissions have also contributed to the move away from the prevailing view that promoting development is an act of charity. Gready's approach is preferred because of its focus on the legitimate right of developing states to acquire the relevant tools to promote their own development.

---

[26] Terminology may vary from 'human rights and development', to 'human rights approach to development', 'rights-based approach to development', and so on.

[27] Andrea Cornwall and Celestine Nyamu-Musembi, 'Putting the "Rights-based Approach" to Development into Perspective' (2004) 25 AmUnivLawRev 1415, 1415–37. The authors clearly state: 'By stipulating an internationally agreed set of norms, backed by international law, it provides a stronger basis for citizens to make claims on their states and for holding states to account for their duties to enhance the access of their citizens to the realisation of their rights.'

[28] ibid 1417.

[29] The UNDP's explanation of the conceptual basis for a rights-based approach in their work is as follows: the central goal of development has been and will be the promotion of human well-being. Given that human rights define and defend human well-being, a rights-based approach to development provides both the conceptual and practical framework for the realization of human rights through the development process, ibid 1426.

[30] Gready (n 1) 737.

[31] ibid 742–44.

Kuosmanen is another author who embraces the rights-based approach to development, although he is of the unique view that the right to development derives from other international legal instruments and therefore does not introduce new demands but instead repackages existing ones within the legal and/or human rights paradigm.[32] There is enough evidence to support the assumption that the current international legal framework can be used to promote innovation in developing states. However, Kuosmanen's submissions are deficient in that, although they focus on developing countries, what was not specified was the particular legal instrument that could be employed to target the promotion of innovation and access to technology. The instrument that can assist in the promotion of innovation and access to technology in developing states is unambiguously the TRIPS Agreement, which contains very pertinent provisions in that regard. Finally, the possibility of devising a new global legal instrument that can promote technology transfer and foster innovation is worth considering.

A notable source is the work of Fink and Maskus, which highlights the positive role of IP rights in stimulating enterprise development and innovation in developing countries.[33] They highlight the crucial part played by the process of transfer of technology to stimulate innovation. Along similar lines, Taylor acknowledges that the transfer of technology has always played a leading role in economic development.[34] Nonetheless, Taylor also notes that international technology transfer is mediated through private markets which are extremely volatile and vulnerable and therefore subject to failure.[35] Concurring with that assertion, Naghavi warns that large corporations may be in a position to determine the markets in which their technology can be disseminated, leaving out those markets which do not meet their requirements in terms of protection of IP rights.[36] In the same context, Maskus observes that countries with inadequate investment climates, weak IP protection, and poor absorptive abilities are unlikely to receive much in the way of inward technology flows.[37] None of these authors provides a solution to the concern raised, but it appears relevant to underline the importance of public intervention to redress this clear sign of market failure and hence push for implementation of relevant IP provisions to address the issue of technology transfer for those countries in need.

---

[32] Kuosmanen (n 2) 303–20.
[33] Keith E Maskus and Carsten Fink (eds), *Intellectual Property and Development: Lessons from Recent Economic Research* (1st edn, World Bank Publications 2005).
[34] M Scott Taylor, 'TRIPS, Trade, and Technology Transfer' (1993) 26 CJE/Revue canadienne d'économie 625, 625–37.
[35] ibid.
[36] Alireza Naghavi, 'Strategic Intellectual Property Rights Policy and North-South Technology Transfer' (2007) 143 RevWorldEcon/Weltwirtschaftliches Archiv 55, 76.
[37] Maskus (n 6) 7.

C. Fragmentation in the Legal Framework on Technology Transfer

Translating the philosophical construct as detailed into a practical tool to promote technology transfer and innovation for the benefit of developing states requires a clear international legal framework. However, as early as the 1960s it was observed that an international legal framework to govern the issue of technology transfer was sorely lacking.[38]

Some attempts were made to establish such an international regime, but without much success. One such case is the UNCTAD Draft International Code of Conduct for the Transfer of Technology. Since the UNCTAD Draft Code was abandoned in 1986, new areas of public interest that were not considered in the negotiations have emerged, such as health, agriculture (protection of new varieties of plants), and climate change.[39] For these new areas, international legal instruments provide for their own separate regimes of technology transfer. Consequently, there are currently around thirty multilateral agreements, half of which are related to global environmental protection, that deal with technology transfer.[40] UNCTAD has identified over eighty international instruments and numerous subregional and bilateral agreements that contain provisions related to technology transfer.[41] Some of the relevant international instruments that contain provisions on technology transfer include: the United Nations Convention on the Law of the Sea of 1982, the Montreal Protocol on Substances that Deplete the Ozone Layer of 1987, the Convention on Biological Diversity (CBD) of 1992, and the United Nations Framework Convention on Climate Change (UNFCCC) of 1992.

Analysis of the sectoral legal regimes established with regard to technology transfer and their main deficiencies, as well as the need to redress them, will be undertaken. Cumulatively, the purpose was to justify the relevance of a harmonized global system on technology transfer.

1. The United Nations Covenant on Economic, Social and Cultural Rights Framework and Technology Transfer

The legal basis that sets out the collective responsibility for the promotion of innovation in developing countries lies in the UN Covenant on Economic, Social

---

[38] Sell (n 4) 99.
[39] Padmashree Sampath and Pedro Roffe, 'Unpacking the International Technology Transfer Debate: Fifty Years and Beyond' (2012) <https://papers.ssrn.com/abstract=2268529> accessed 20 June 2021.
[40] Ma (n 7) 77.
[41] UNCTAD, *Compendium of International Arrangements on Transfer of Technology: Relevant Provisions in Selected International Arrangements Pertaining to Transfer of Technology* (United Nations 2001).

and Cultural Rights (CESCR).[42] In Article 15(1)(b) the CESCR provides for the right to the enjoyment of the benefits of scientific progress and its applications. Article 15(2) emphasizes that the steps to be taken by states parties to achieve the full realization of this right shall include those necessary for the development and diffusion of science. It must also be highlighted that the Covenant is emphatic in Article 2(2) that the rights enunciated must be exercised without discrimination of any kind as to national or social origin. Based on these provisions, it follows that no one, regardless of the state of origin, should be deprived of the right to enjoy the benefits of scientific and technological progress achieved by humankind. However, in the current global context, private corporations are the major rights-holders of scientific and technological output and proprietary rights encapsulated in the IP system and, very often, it is they who control access to this vital information.[43] This poses serious challenges to developing states seeking to facilitate access to the fruits of technological progress for the benefit of their citizens.[44] Indeed, the diffusion of science as required by Article 15(2) of the Covenant can only take place through the mechanism of voluntary technology transfer. This is the major challenge that has raised controversy and heated debates on the issue of technology transfer.

2. The UNCTAD Draft International Code of Conduct for the Transfer of Technology

The issue of technology transfer was first raised in the national laws of Latin American states. It was later transposed to the negotiations for the establishment of the New International Economic Order (NIEO), which for the first time directly linked technology transfer to economic development.[45] The Declaration on the establishment of the NIEO posited that developing states should be given access to the achievements of modern science and technology, and that the transfer of technology and the creation of indigenous technology should specifically be promoted for the benefit of developing states.[46] Consequently, the Programme of Action for the establishment of the NIEO demanded the adoption of an International Code of Conduct on the Transfer of Technology.[47] UNCTAD developed a Draft International Code of Conduct for the Transfer of Technology to remove

---

[42] International Covenant on Economic, Social and Cultural Rights (adopted 6 December 1966, entered into force 3 January 1976) GA Res 2200A (XXI), 999 UNTS 3 (hereafter ICESCR).

[43] Saon Ray, 'Technology Transfer and Technology Policy in a Developing Country' (2012) 46 JDA 371, 373.

[44] Philippe Cullet, 'Human Rights and Intellectual Property Protection in the TRIPS Era' (2007) 29 HRQ 403, 408.

[45] Sell (n 4) 199.

[46] United Nations Resolution on the establishment of the New International Economic Order (adopted 1 May 1974) Doc A/RES/S-6/3201, para 4(p).

[47] United Nations Programme of Action on the Establishment of a New International Economic Order (adopted 1 May 1974) Doc A/RES/S-6/3202 (hereafter NEIO).

constraints imposed due to domination of the international technology market by multinationals on the acquisition of technology by developing states.[48] The Draft Code proposed liberalization of trade in technology and the introduction of guidelines on the terms and conditions of transfer of technology to developing states.[49]

Notwithstanding protracted negotiations that lasted from 1976 to 1985, the Draft Code was never adopted by the General Assembly due to a lack of consensus.[50] The primary reason why the UNCTAD Draft Code failed to be adopted is the fact that developing states consistently pushed for mandatory effect of the instrument, while developed states were instead seeking non-binding guidelines or some sort of flexible standards to be observed by the companies involved in technology transfer.[51] Secondly, the Draft Code failed to be adopted due to profound divergence of the positions between developing and developed states with regard to its content, especially Chapter 4, which was related to restrictive business practices. The point of discord and dissension was the criteria to be used to outlaw the practices: developed states considered that a practice must be prohibited if, cumulatively, it had an impact on competition by being unduly restrictive and adversely affecting international technology transfer, whereas developing states considered that those elements were not cumulative. Furthermore, the developed states were only concerned with the practice if this had an impact on their economic and technological development, and it was only in those exceptional cases that they would consider outlawing it.[52] Thirdly, there was no consensus regarding the strength and nature of the condemnation of objectionable practices. Developing states had been pushing for automatic prohibition of designated objectionable practices, while developed states required initial administrative scrutiny of practices based on the 'rule of reason' before such a decision was taken.[53] Even with regard to the list of unlawful practices to be prohibited, which ranged from fourteen to twenty, consensus was still lacking.[54] Lastly, until the negotiations came to a standstill, consensus was yet to be achieved with regard to Chapter 9 related to the applicable law, although some progress had been made in this regard.[55]

In view of the failure to adopt the proposed UNCTAD International Code of Conduct for the Transfer of Technology, there is no unified or harmonized legal

---

[48] UNCTAD (n 41).
[49] UNCTAD, *Transfer of Technology* (United Nations 2001).
[50] Peter Gottschalk, 'Technology Transfer and Benefit Sharing under the Biodiversity Convention' in Hans Henrik Lidgard and others (eds), *Sustainable Technology Transfer: A Guide to Global Aid & Trade Development* (Kluwer Law International 2012) 200.
[51] In the proposed text of the preamble by the developing countries it was clearly stated that 'an internationally legally binding instrument is the only form capable of effectively regulating the transfer of technology', hence such instrument was vividly recommended for adoption. See also Michael Blakeney, *Legal Aspects of the Transfer of Technology to Developing Countries* (ESC Pub 1989) 134–35.
[52] ibid 142–43.
[53] ibid 143.
[54] ibid 144–50.
[55] ibid 158–61.

international regime to govern the transfer of technology in the world. Instead, several sectoral regimes of technology transfer are currently in place, denoting the typical situation of fragmentation that ought to be addressed.

3. The United Nations Convention on the Law of the Sea and Transfer of Technology

The United Nations Convention on the Law of the Sea was adopted on 10 December 1982. The Treaty is a comprehensive effort to regulate the world's oceans and includes provisions on technology transfer. In this regard, the Treaty creates an International Seabed Authority to regulate the mining of strategic minerals from the deep seabed. The main provisions on technology transfer are contained in Article 144, which establishes that the International Seabed Authority shall take measures in accordance with the Convention to acquire technology and scientific knowledge relating to activities in the deep seabed and to promote and encourage the transfer to developing states of such technology and scientific knowledge so that all states parties shall benefit therefrom. The Treaty contains an Annex also entitled 'Transfer of Technology' (Annex III) with detailed instructions dealing with mandatory transfer of marine technology. The Treaty already provided a framework for the systematic transfer of marine technology from the industrial world to the developing states. However, with specific regard to the seabed mining associated technology, the Treaty sought to make its transfer mandatory.[56]

Although the industrialized states agreed, in principle, to the concept that transfer of technology is an acceptable political and philosophical idea, they argued that the Treaty should not contain provisions for the mandatory transfer of private technologies.[57] The insertion of the specific clause that dictates mandatory transfer of technology is one of the main reasons behind the reluctance on the part of the United States (US) and other western states to sign and ratify the Treaty. Conversely, developing states favour a mandatory regulation of the flow of technologies as an opportunity to share the wealth, prosperity, and property that are monopolized by western states.[58]

The mandatory obligation to transfer technology established by Article 144 seems to have been watered down by the United Nations Agreement Relating to the Implementation of Part XI of the United Nations Convention on the Law of the Sea of 1994, which states in section 5 that companies and developing states wishing to obtain deep seabed mining technology are obliged to seek to obtain

---

[56] James Stavridis, 'Marine Technology Transfer and the Law of the Sea' (1983) 36 NWCR 38, 41.

[57] ibid 41–42; DH Anderson, 'Efforts to Ensure Universal Participation in the United Nations Convention on the Law of the Sea' (1993) 42 ICLQ 654, 658.

[58] Stavridis (n 56) 42.

such technology on fair and reasonable commercial terms and conditions on the open market, or through joint-venture arrangements. In circumstances where it is not possible to obtain technology on fair and reasonable commercial terms, the Treaty calls for cooperation to facilitate the acquisition of this technology, on condition that this arrangement is consistent with the effective protection of IP rights.[59] Since 1994, some western states have acceded to the Treaty, such as: Australia, Austria, Canada, Denmark, France, Germany, Italy, Netherlands, Norway, Portugal, Russian Federation, Spain, Sweden, Switzerland, and the United Kingdom. However, the US is still conspicuous in its absence from membership of this important arrangement.

4. The United Nations Framework Convention on Climate Change Regime of Technology Transfer and the Kyoto Protocol

The UNFCCC of 1992 established a comprehensive regime to promote technology transfer and a Technology Mechanism on climate change.[60] The UNFCCC recognizes the necessity for rapid development by way of the transfer and use of environmentally sound technologies that can effectively reduce greenhouse gas emissions in various provisions, namely Articles 4.1, 4.3, 4.5, 4.7, 4.8, and 4.9. Specifically, Article 4.5 tasks developed states with 'taking all practicable steps to promote, facilitate and finance, as appropriate, the transfer of, or access to, environmentally sound technologies and know-how' to other state parties, particularly developing states parties, to enable them to implement the provisions of the Convention; and to 'support the development and enhancement of endogenous capacities and technologies' of developing states parties in the process. In addition, any other parties and organizations that find themselves in a position to do so should assist in facilitating the transfer of such technologies.

As an ancillary treaty to the UNFCCC, the Kyoto Protocol contains provisions directly concerning technology transfer in a number of its articles and these impose obligations on developed states to transfer environmentally sound technologies to developing states. Article 2 compels developed states to transfer technology for purposes of performing their obligations of implementing and elaborating policies and measures on increased use of environmentally sound technologies. Article 10(C) requires parties to the Protocol to cooperate in the promotion of

---

[59] Annex 5 of the United Nations Agreement Relating to the Implementation of Part XI of the United Nations Convention on the Law of the Sea (adopted 10 December 1982, entered into force 16 November 1994) 1836 UNTS (hereafter Part XI Implementation Agreement).

[60] The Technology Mechanism comprises an implementing body, the Climate Technology Centre and Network, and a policy body, the Technology Executive Committee, composed by twenty technology experts representing both developed and developing states whose role it is analyse issues and to provide policy recommendations that support state efforts to enhance climate technology development and transfer.

effective modalities for the development, application, and diffusion of environmentally sound technologies. This encompasses the duty to take all practicable steps to promote, facilitate, and finance, as appropriate, the transfer of, or access to, environmentally sound technologies, know-how, practices, and processes pertinent to climate change, in particular to developing states, including the formulation of policies and programmes for the effective transfer of environmentally sound technologies that are publicly owned or in the public domain and the creation of an enabling environment for the private sector, to promote and enhance the transfer of, and access to, environmentally sound technologies.

Although the Kyoto Protocol appears to have provisions that are legally binding, the terminology used, such as 'encourage', 'facilitate', 'promote', 'exert efforts', and 'assist', entails a flexible meaning of each concept and therefore does not impose definite and binding commitments on states from whom compliance can be expected. Therefore, the provisions of the Protocol are regarded as soft law and their effectiveness depends on the existing domestic laws and regulations of each party. Nevertheless, it appears that the Protocol has contributed to some extent to facilitating technology transfer to developing states.[61]

5. The Convention on Biological Diversity and Transfer of Technology

The Convention on Biological Diversity (CBD) of 1992 also contains several provisions related to the transfer of technology. Central to the issue, however, is Article 16, which states that each contracting party undertakes to provide and/or facilitate access and transfer to other contracting parties, of technologies that are relevant to the conservation and sustainable use of biological diversity. Article 23 of the Nagoya Protocol on Access to Genetic Resources and the Fair and Equitable Sharing of Benefits Arising from their Utilization reiterates these obligations by stating that the parties shall undertake to promote and encourage access to technology by, and transfer of technology to, developing states, in order to enable the development and strengthening of a sound and viable technological and scientific base for the attainment of the objectives of the Convention and the Protocol. Having said that, the main legal challenge to these provisions concerns the limitation of state efforts in sharing technology that is in the hands of private actors, particularly when protected by IP rights. Therefore, the Protocol merely encourages member states to adopt domestic measures that provide, or at least facilitate, access to technologies that make use of genetic resources, including biotechnologies, as well as technologies that are relevant for conservation and sustainable use.[62] Based

---

[61] Ma (n 7) 79.
[62] Elisa Morgera, Elsa Tsioumani, and Matthias Buck, *Unraveling the Nagoya Protocol: A Commentary on the Nagoya Protocol on Access and Benefit-Sharing to the Convention on Biological Diversity* (Brill Nijhoff 2014) 314–21.

on Article 16(2), this has to be arranged under fair and most favourable terms, including on concessional and preferential terms if mutually agreed, where developing states are concerned.

6. The Vienna Convention and the Montreal Protocol on Substances that Deplete the Ozone Layer and Technology Transfer

The Vienna Convention on Substances that Deplete the Ozone Layer was adopted in 1985. The parties to this Convention agreed to prevent further deterioration of the ozone layer and to promote the exchange of information and the coordination of legal and administrative measures to this end. The Convention constitutes a framework without specifying the obligations. In this regard, Article 8 of the Convention defers that task to subsequent protocols to be adopted by the parties.

The Montreal Protocol on Substances that Deplete the Ozone Layer of 1987 was adopted under the auspices of the United Nations Environment Programme (UNEP) and sought to establish a more detailed framework for limiting production of chlorofluorocarbons (CFCs) while simultaneously freezing halon production. The Montreal Protocol also recognizes the special need of developing states for financial and technological assistance. Therefore, the parties agreed to develop appropriate funding mechanisms to enable developing states to join the Protocol and to facilitate their access to scientific information, research results, and training. With regard to technology transfer, the London Amendment to the Protocol adopted in 1990 introduced Article 10A of the Protocol, which establishes that each party shall take every practicable step, consistent with the programmes supported by the financial mechanism, to ensure that the best available and most environmentally safe substitutes and related technologies are expeditiously transferred to parties under fair and most favourable conditions. Concerns raised by developing states were also addressed through the new formulation of Article 5(5), which establishes that their capacity to fulfil their obligations (control measures) depends upon the effective implementation of the financial cooperation provisions as elucidated in Article 10 and the transfer of technology as provided for in Article 10A.[63]

The new Article 10A and the new formulation of Article 5 constitute a compromise that emerged after lengthy debate, where developing states were insisting that they would only comply with the Protocol if financial support was provided by developed states and where technology was transferred. Unsurprisingly, developed states refused to accept that technology was IP owned by private industry. The new provisions seem to reconcile the two positions.

---

[63] Hermann Ott, 'The New Montreal Protocol: A Small Step for the Protection of the Ozone Layer, a Big Step for International Law and Relations' (1991) 24 VRÜ/WCL 188, 200.

D. Proposed Reforms on the International Framework for Regulating Technology Transfer

1. The Proposed Multilateral Agreement on Access to Basic Science and Technology under the WTO Framework

Cognizant of the lack of a legal instrument to promote the dissemination of technological information and technology transfer and innovation, in 2004 Keith Maskus tabled a proposal for a Multilateral Agreement on Access to Basic Science and Technology (ABST).[64] The premise of the proposal by Maskus was to expand the availability of technologies in the public domain or promote access at an affordable cost. Therefore, Maskus proposed that an agreement at the World Trade Organization (WTO) would be negotiated to place outputs of publicly funded research into the public domain, or devise mechanisms to make available basic technological information at a modest cost. The framework, as proposed, also sought to balance protection of IP rights in commercial technologies while preserving and enhancing the global commons in science and technology and setting out a public mechanism for increasing the international flow of technical information for developing states.[65]

The proposal was first developed by John Barton in 2003[66] and further consolidated by Maskus and Barton in 2004.[67] According to the proponents, the Agreement would find its foundations in open access with the objective of coordinating and promoting movement of research projects and scientific personnel as well as disseminating basic research results.[68] However, it is worth recalling that when Barton and Maskus suggested the ABST arrangement, they had envisaged that it would be developed under the WTO.[69] Naturally, this is because the WTO is recognized as the custodian of the TRIPS Agreement, which is a multilateral arrangement that disciplines free riding by promoting the enforcement of IP rights, thereby curtailing any illicit flow of protected technologies or scientific results. Indeed, the WTO has a specific responsibility to promote the transfer of technology as a result of the provisions of Articles 7 and 66.2 of the TRIPS Agreement.[70] Apart

---

[64] Maskus (n 6) 38.
[65] ibid.
[66] John Barton, 'Preserving the Global Scientific and Technological Commons' (2003) paper presented to Science and Technology Diplomacy Initiative and the ICTSD-UNCTAD Project on IPRs and Sustainable Development Policy Dialogue on a Proposal for an International Science and Technology Treaty, 11 April 2003.
[67] John Barton and Keith Maskus, 'Economic Perspectives on a Multilateral Agreement on Open Access to Basic Science and Technology' (2004) 1 SCRIPT-ed 369–87.
[68] ibid 370.
[69] ibid 379.
[70] Marrakesh Agreement Establishing the World Trade Organization, Annex 1C, 1869 UNTS 299 (opened for signature 15 April 1994, entered into force 1 January 1995) containing the Agreement on Trade-Related Aspects of Intellectual Property Rights (hereafter TRIPS Agreement)

from the TRIPS Agreement, the WTO also manages agreements governing other topics that are related to the transfer of scientific results such as subsidies, standards, and trade in services.[71]

According to the proponents of the Agreement, the WTO is also a forum where negotiating processes occur, thus permitting trade-offs in concessions across sectors and functional agreements. Accordingly, the WTO is a focal point for the strengthening of national constituencies seeking the benefits of multilateral agreements. The WTO system also contains many of the essential principles that could be applied to ABST and ensure that it is effective.[72] Particular tools that give effect to this objective include the Special and Differential Treatment provisions and the Dispute Settlement Mechanism (DSM).

The Special and Differential Treatment provisions are contained in various WTO Agreements and confer on developing states special rights which enable developed states to treat them more favourably than other WTO members.[73] Developing states benefited from an initial four-year transitional period for protected technologies. An additional transitional period of five years was available until 1 January 2005 with respect to technologies that were not patentable under domestic laws prior to the date that TRIPS applied to that member. As pertains specifically to LDCs, the TRIPS preamble and Article 8(1) recognize the need for differential treatment to allow LDCs to create a sound technological base. Indeed, LDCs are not required to apply the provisions of the TRIPS Agreement, save for Articles 3, 4, and 5, until 1 July 2034 or when they cease to be an LDC (whichever occurs first). This extension is the third granted to the LDCs, with the first granted in 2005 and the second, which expired on 1 July 2021, granted in 2013. In addition, there is a pharmaceutical transition period until 1 January 2033 or when an LDC ceases to be an LDC, whichever occurs first. Under this transition period, an LDC does not have to issue pharmaceutical patents. Moreover, LDCs benefit from the possibility of issuing compulsory licences.

With regard to technology transfer and differential treatment for developing states and LDCs, it is required that all WTO members must safeguard the trade interests of developing states; they must support developing states to build the capacity to carry out WTO work, handle disputes, and implement technical standards. Specific provisions relating to LDC members, such as Article 66.2, were crafted to cater for their special needs.

The Dispute Settlement Mechanism offers a format for arbitrating and settling disputes arising between states and/or their governments.[74] It is submitted that this element is of fundamental importance in that the ABST mechanism's sustainability

---

[71] Barton and Maskus (n 67) 379.
[72] ibid.
[73] WTO, 'Special and Differential Treatment Provisions' <www.wto.org/english/tratop_e/devel_e/dev_special_differential_provisions_e.htm> accessed 9 August 2019.
[74] Barton and Maskus (n 67) 379.

is undoubtedly anchored to an effective enforcement mechanism. Very often, transfer of technology does not happen voluntarily, and states may need an enforcement tool to pursue their quest to access technologies against other states. The DSM offers the requisite recognized and authoritative avenue to settle disputes between governments, albeit that it may be improved, facilitating the process of appointing Appellate Body members.

When Barton and Maskus presented the proposal of an ABST in 2004 they were confident that this would revolutionize the flow of technology to developing states. However, they were mindful of the challenges that a legally binding treaty would face in receiving the necessary political support and therefore they went as far as to suggest a compromise by lessening some of the requirements.[75] The main area identified is related to the granting of patents for basic research tools and software that is common practice for example in the US but that could be challenged in other states. The proponents therefore recommended a compromise in which developed states which protect that knowledge could limit the breadth of the claims in the patent applications and further adopt licensing manuals and fee schedules that discriminate in favour of users in developing states and educational institutions. For their part, developing states could avoid adopting weak eligibility standards for patents and broad claims but could still maintain research and education exemptions for the use of patented knowledge.[76]

Although relevant, the ABST proposal has never been developed into a concrete instrument and was never considered by the WTO. One of the challenges in locating the ABST under the WTO is related to the fact that it focuses on the results of publicly funded research and aims at enhancing the transfer of basic technological information. In view of this, it is submitted that although the proponents have expressed preference for seeing such an Agreement adopted under the WTO, the outputs of publicly funded research are not of relevance to this organization, which focuses on trade-related issues. Arguably, this aspect may have induced the WTO to ignore the proposal due to its irrelevance to the objectives of the organization.

2. The Proposed Implementation of ABST under the WIPO Development Agenda

An attempt to implement the framework proposed by Barton and Maskus was instead undertaken in the context of the proposal for the establishment of a Development Agenda for the WIPO that was presented in 2004 by requesting the

---

[75] Maskus (n 6) 38.
[76] Barton and Maskus (n 67) 379.

adoption of a Treaty on ABST and the inclusion of provisions on transfer of technology in the treaties under negotiation in the WIPO.[77]

The Development Agenda for WIPO was adopted by the member states of WIPO during the 43rd session of the General Assembly that was held from 24 September to 3 October 2007.[78] The document comprised forty-five recommendations, structured into six clusters.[79] The General Assembly also established the Committee on Development and Intellectual Property (CDIP) mainly to ensure the implementation of the adopted recommendations. Cluster C of the Development Agenda is the most relevant to this book as it is dedicated to 'Technology Transfer, Information and Communication Technologies and Access to Knowledge'. In addition, Cluster F, provides a linkage with the TRIPS Agreement by suggesting that the enforcement of IP rights should contribute to the promotion of technological innovation and to the transfer and dissemination of technology in accordance with Article 7 of the TRIPS Agreement.

The final—and streamlined—'WIPO Agenda for Development' adopted by the WIPO Assemblies in 2007 does not include reference to a Treaty on ABST in order to promote transfer of technology and foster innovation. A softer approach with regard to transfer of technology for the benefit of developing states and LDCs was included in recommendations 25, 26, 28, 29, 30, and 31 of the Agenda for Development. Those recommendations merely call for the need to explore IP-related initiatives, measures, and policies that can promote transfer of technology, access to knowledge, and technological information and the promotion of debates surrounding the topic. The recommendations do not constitute any legally binding tool; hence WIPO or developed states are not compelled, nor even encouraged, to take measures that will effectively promote technology transfer to the states in need.[80]

The WIPO endeavoured to implement the Agenda for Development directive by developing a project entitled 'Innovation and Technology Transfer Support Structure for National Institutions' to stimulate local innovation and technology transfer in developing states (especially LDCs) through establishing and improving

---

[77] 'Proposal by Argentina and Brazil for the Establishment of a Development Agenda for WIPO' Doc WO/GA/31/11 <www.wipo.int/edocs/mdocs/govbody/en/wo_ga_31/wo_ga_31_11.pdf> accessed 8 July 2018; Andrew C Michaels, 'International Technology Transfer and TRIPS Article 66.2: Can Global Administrative Law Help Least-Developed Countries Get What They Bargained For?' (2009) 41 GJIL 258.

[78] 'Development Agenda for WIPO' <www.wipo.int/ip-development/en/agenda/index.html> accessed 8 July 2018.

[79] The six clusters are as follows: A) Technical assistance and capacity-building; B) Norm-setting, flexibilities, public policy, and public domain; C) Technology transfer, information and communication technologies (ICT), and access to knowledge; D) Assessment, evaluation, and impact studies; E) Institutional matters including mandate and governance; F) Other issues.

[80] Keith E Maskus, 'The WIPO Development Agenda' in Neil Weinstock Netanel (ed), *The Development Agenda: Global Intellectual Property and Developing Countries* (Oxford University Press 2008) 167.

the necessary infrastructure (both legal and organizational) and developing professional skills for the effective use of the IP system in the area of innovation and technology transfer.[81] However, this project was very limited in scope as it only envisaged capacitating academic and research institutions in IP and technology management and the establishment of a digital repository on the WIPO's website for relevant documents on the subject matter.[82]

Based on the facts illustrated, situating the technology transfer treaty within the WIPO is challenging, primarily because of its limited scope. Therefore, it is submitted that even if a stronger stance had been taken under the WIPO Agenda for Development process to address the issue of technology transfer, it would have been limited to addressing issues related to the intersection between IP and technology transfer. However, technology transfer is a cross-cutting issue that goes beyond IP and involves trade, the environment, and so on and so forth. Therefore, establishing a legal and institutional framework in the WIPO context would not have been enough; at most it would have been another manifestation of the fragmentation phenomenon. In view of that, this book does not encourage an isolated initiative within the WIPO but instead advocates for collective action with all other institutions that include policy areas involving the transfer of technologies. This is the reason why the WTO context seems to be more prone to mobilizing all relevant sectors on the issue of technology transfer in view of the unifying factor that trade represents.

Notwithstanding this setback, the idea of a treaty on ABST is not yet dead. The Report on the WIPO expert forum on international technology transfer presented to the 15th session of the Committee on Development and Intellectual Property (CDIP) still insists on the need for the establishment of such an international treaty.[83] It appears that there is merit in that proposal because the important role played by technology transfer in promoting development, especially for the LDCs, can only be sustainable if it is backed by a specific international legal instrument, as suggested by the rights-based approach to development.

It was accordingly submitted that there is no doubt that the current fragmentation in the legal framework related to transfer of technology is a serious weakness. Therefore, although the UNCTAD Draft International Code of Conduct for the Transfer of Technology never saw the light of day and the proposed ABST was shot down under the WIPO Agenda for Development and the WTO, the quest by developing countries for a binding legal instrument to promote technology transfer cannot be frustrated because it remains a noble objective. For African LDCs,

---

[81] WIPO, 'Innovation and Technology Transfer Support Structure for National Institutions (Recommendation 10)' (2010) CDIP/3/INF/2/STUDY/VII/INF/1 <https://dacatalogue.wipo.int/projects/DA_10_03> accessed 29 October 2023.
[82] ibid.
[83] WIPO, 'Report on the WIPO Expert Forum on International Technology Transfer' (2015) CDIP/15/5 <www.wipo.int/meetings/en/doc_details.jsp?doc_id=298371> accessed 26 August 2018.

pushing for a mechanism to promote the transfer of technology in future negotiations in global fora such as the TRIPS Council and WIPO norm-setting debates, claiming the establishment of an international legal framework could be a valid weapon to be pursued.

3. Proposal to Establish an International Technology Transfer Agreement under the WTO

A second proposal suggests taking advantage of the effectiveness of the Kyoto Protocol to the UNFCCC to categorically establish the legal system of international technology transfer.[84] The Kyoto Protocol is an international agreement linked to the UNFCCC adopted in Kyoto, Japan, on 11 December 1997 and which entered into force on 16 February 2005.[85] Several provisions of the UNFCCC urge its member states to develop, use, and transfer environmentally sound technologies to poor states. It also exhorts its members to fund and support the development and enhancement of endogenous capacities and technologies of developing states parties.[86] More precise and comprehensive on the matter is Article 10(C) of the Kyoto Protocol, which states that members must:

> cooperate in the promotion of effective modalities for the development, application and diffusion of, and take all practicable steps to promote, facilitate and finance, as appropriate, the transfer of, or access to, environmentally sound technologies, know-how, practices and processes pertinent to climate change, in particular to developing countries, including the formulation of policies and programmes for the effective transfer of environmentally sound technologies that are publicly owned or in the public domain and the creation of an enabling environment for the private sector, to promote and enhance the transfer of, and access to, environmentally sound technologies.[87]

The main challenge of the current regime is that the effectiveness of these instruments is entirely dependent on the domestic laws and regulations of each party because the international regime lacks a clear definition, is non-binding in character, its obligations are vague, and the compliance mechanisms are ineffective.[88]

---

[84] Ma (n 7) 82–85.
[85] Kyoto Protocol to the United Nations Framework Convention on Climate Change (opened for signature 11 December 1997, entered into force 16 February 2005) 2303 UNTS 162 (hereafter Kyoto Protocol).
[86] Some of the relevant provisions on transfer of technologies included in the UNFCCC are Articles 2, 3.14, and 4.5, while art 10(C) of the Kyoto Protocol is key.
[87] The text of the Kyoto Protocol is available at <https://unfccc.int/sites/default/files/kpeng.pdf> accessed 13 June 2019.
[88] Ma (n 7) 82.

Considering also that those technologies are owned by private entities and are protected by IP rights, it seems improbable that an owner will voluntarily abdicate his rights simply for the sake of making a contribution to diffusing and transferring environmentally sound technologies for the benefit of humanity, unless he is required to do so by law.[89]

This scenario has motivated some scholars to suggest that the technology transfer provisions of the Kyoto Protocol and other international treaties could be assembled to compose a multilateral, legally binding agreement—the International Transfer of Technology Agreement. The ITT should have the format of a multilateral agreement adopted by the Ministerial Conference with the same status as that of the General Agreement on Trade in Services (GATS), General Agreement on Tariffs and Trade (GATT), Agreement on Trade-Related Investment Measures (TRIMS), and TRIPS under the WTO, with a focus on promoting the transfer of environmentally sound technologies.[90] The ITT is particularly relevant because it addresses, in a specific way, the important issue of implementation. Indeed, capitalizing on the trade context of the WTO, the Agreement could contain an obligation for the member states to enact national legislation that domesticates it, hence making it easier to persuade the private sector to participate in the transfer of technologies for the benefit of developing states. Violation of these obligations would trigger the Dispute Settlement Mechanism of WTO, thus causing the ITT to be enforceable.[91]

Proponents of the ITT also further elaborate on how the new regime could persuade the private sector to actively provide access to their owned and protected technologies. The private sector depends on governments to stimulate demand and supply of environmentally sound technologies, and since the implementation of the ITT would be made operational through national policies and legislation, governments—especially from the developed world—would be in a better position to influence those factors.[92] At the international level, a Monitoring Mechanism would be created, according to the proponents, to assess whether the member states are implementing the Agreement by enacting the necessary policies, legal frameworks, and incentives.[93]

This proposal is persuasive and contains several elements that this book is inclined to subscribe to. However, the proposal falls short in view of the limitation of its scope, particularly due to the focus of the proponents on global environmental protection. It is also noteworthy that, of the twenty-eight multilateral agreements that address the issue of technology transfer, about half are related to environmental protection and this may have had a significant bearing on the submissions

---

[89] ibid 88.
[90] ibid 92.
[91] ibid.
[92] ibid 90.
[93] ibid 93.

of the proponents. Thus, the proposed ITT focus on promoting transfer of environmentally sound technologies.[94] No compelling reasons were found for why the treaty should focus only on environmental technologies, to the exclusion of any other technology that would be relevant to uplifting developing states. That approach risks perpetuating the adoption of new sectoral solutions on the same issue, which will ultimately lead to fragmentation, a scenario that is indeed undesirable. Furthermore, the designation of the legal instrument was not a major concern of the proponents, who only identified it as a legal system with the objective of promoting ITT. This is not descriptive of the type, legal nature, and content of the proposed future instrument. This book leverages on the strengths of this important proposal and expands upon it to cover all international technology transfer regimes. Enriching and expanding this proposal appears, for purposes of this book, to be a unique opportunity to finally provide a definitive solution to the lack of a unitary framework to govern the international technology transfer regime that the world has been seeking for the last fifty years. Chapter 6 of this book therefore elaborates further on the proposals.

E. Conclusion

The failure of the UNCTAD Draft Code in 1986 has left a space that is yet to be filled with regard to the establishment of an international legal regime of technology transfer. In the meantime, the situation was exacerbated by the emergence of new areas of public interest with a major bearing on the matter, such as health, agriculture (protection of new varieties of plants), and climate change.[95] The solution that was found for these new areas was to develop sectoral international legal instruments that provide for their own separate regimes of technology transfer. This has further fuelled the fragmentation of the legal regime that currently boasts around thirty international agreements that deal disjointedly with technology transfer.

In view of this, the proposal for the adoption of a Multilateral Agreement on Access to Basic Science and Technology that was tabled by Maskus initially appeared to fill the vacuum.[96] However, the reluctance of either the WTO or the WIPO to embrace it perpetuated the lack of a harmonized regime and the two organizations are yet to table a concrete solution to this gap in the international legal framework. Nonetheless, there remain a few challenges to the establishment of an internationally agreed upon legal instrument for the promotion of technology transfer and these are mainly related to the scope of the potential instrument, its binding nature, and the most adequate institutional framework that can host it.

---

[94] ibid 75, 77, 90.
[95] Sampath and Roffe (n 39) 49.
[96] Maskus (n 6) 38.

These challenges should not frustrate developing states in their quest for such a binding legal instrument to promote technology transfer.

The latest proposal for the establishment of an International Transfer of Technology Agreement based on some of the features of the Kyoto Protocol and other international treaties seems to address some of the challenges identified.[97] Indeed, such an agreement should take the form of a multilateral agreement with the same status as the GATS, GATT, TRIMS, and TRIPS, situated within the WTO. However, one of the shortcomings of this proposal is related to its scope, which is limited to promoting transfer of environmentally sound technologies. No compelling reasons were found to explain why such an important and necessary instrument should be limited to some technologies to the exclusion of all others. Therefore, Chapter 6 will explore the possible development of an omnibus treaty on technology transfer taking into account its cross-cutting nature and the aspirations of developing states, especially in Africa.

Although the adoption of an international legal instrument on international technology transfer is the ideal solution, it is also prudent to consider the intricacies involved in achieving such a result within a short period of time. Chapter 5 therefore explores the possibility of improving the already existing, although limited, regime of technology transfer that was established by Article 66.2 of TRIPS in order to enable, in the interim, some flows of technology to LDCs.

---

[97] Ma (n 7) 92.

5
Maximizing the Use of the TRIPS Agreement to Promote Technology Transfer and Innovation in Africa

A. Introduction

In Chapter 4, a bold proposal was made for the adoption of a new international legal instrument to promote the transfer of technology. Whereas it was submitted that such a proposed treaty is the most appropriate tool to grant enhanced legal certainty on the issue of technology transfer at the global level, it was also conceded that negotiating a treaty is tricky; it is a lengthy process and may entail unpredictable results. It is to be recalled, in this context, that the United Nations Conference on Trade and Development (UNCTAD) Draft International Code of Conduct for the Transfer of Technology took a decade of negotiations and was finally abandoned in 1985. Therefore, although the inability of Article 66.2 to trigger flows of technology to Least Developed Countries (LDCs) has been widely documented, some improvements are possible and substantial results are achievable.

Philosophically, it is to be recalled that Kuosmanen defended the fact that the right to development can be extracted from existing international legal instruments.[1] Based on that assumption, the current international legal framework that can be used to promote innovation in developing states is the Agreement on Trade-Related Aspects of Intellectual Property Rights (TRIPS Agreement). Accordingly, this chapter tables more practical proposals to improve the implementation of the existing Article 66.2 and thus achieve some of its objectives.

While the text of Article 66.2 seems straightforward, paradoxically, since its inception numerous issues have been raised pertaining to its interpretation and implementation. Consequently, the objective of promoting transfer of technology to LDCs through this provision has not been achieved. The challenges in the interpretation of Article 66.2 include, among others, the identification of the duty-bearers of the obligations arising from the provision; the scope, including the clear meaning and type of the incentives to be provided to companies in their territories; the purpose of the provision; the indicators of achievement of the goals pursued;

---

[1] Jaakko Kuosmanen, 'Repackaging Human Rights: On the Justification and the Function of the Right to Development' (2015) 11 JGE 303, 303–20.

and the precise beneficiaries. Negotiators at the TRIPS Council Sessions and academics have been struggling to clarify some of the concepts contained in this provision to enable its effective implementation. This chapter contributes to the ensuing debate in an attempt to provide some clarity and certainty and has been structured accordingly.

After the introduction, a section is devoted specifically to clarifying the challenging concepts contained in Article 66.2. The chapter then examines how best to improve the reporting mechanism; monitoring of the implementation of Article 66.2 from the perspective of LDCs; and the potential institutional framework that can effectively support implementation of the provision. Finally, the enforcement mechanism of Article 66.2 is discussed with a focus on the Dispute Settlement Mechanism (DSM). These discussions are then used to draw conclusions. Essentially, the chapter emphasizes that the current debates over the implementation of Article 66.2 centre on the effectiveness of the incentives adopted by developed states to assist LDCs in building a sound and viable technological base for their development. Given this, the recommendations provided in this chapter regarding the reporting mechanism, monitoring of implementation of the provision, institutional arrangements, and enforcement address these shortcomings.

B. Clarifying Article 66.2 of the TRIPS Agreement on Technology Transfer

1. Overview of Article 66.2

Article 66.2 of TRIPS establishes that developed member states shall provide incentives to companies and institutions in their territories for the purpose of promoting and encouraging technology transfer to least-developed member states in order to enable them to create a sound and viable technological base.[2] From the text of this provision, the scope is clearly defined: it consists of placing the responsibility squarely onto the governments of developed states to provide incentives to companies and institutions that are located in their territories. Therefore, the provision does not create an obligation to transfer technology but is limited to assigning a duty on developed states to provide incentives as an enabling factor to promote the process of technology transfer.[3] Clearly it is expected that these incentives must be sufficiently compelling to trigger the dynamics related to the process.

---

[2] Marrakesh Agreement Establishing the World Trade Organization, Annex 1C, 1869 UNTS 299 (opened for signature 15 April 1994, entered into force 1 January 1995) containing the Agreement on Trade-Related Aspects of Intellectual Property Rights

[3] Jayashree Watal and Leticia Caminero, 'Least-Developed Countries, Transfer of Technology and the TRIPS Agreement' 5–6 <www.wto.org/english/res_e/reser_e/ersd201801_e.pdf> accessed 24 April 2018.

The main challenge that is posed with regard to Article 66.2 relates to the question of the effectiveness of these incentives.[4]

Although the text of the provision appears straightforward, challenges have arisen in its implementation, perhaps attributed to ambiguities in the terms embedded in the Treaty.[5] These ambiguities relate to aspects such as clear identification of the duty-bearers of the obligation, or in other words, whether it is governments of developed states or the private sector that owns the technologies; the notion of technology transfer; the type of incentives that must be established by developed states to entice their companies to transfer technology; the type of technology involved; and whether that technology falls within the ambit of intellectual property (IP) rights.[6]

This assessment sheds light on some of the contentious issues raised and proposes improvements in the interpretation of the provision, including institutional arrangements and the streamlining of the processes with a view to drawing increased benefits for the beneficiary LDCs.

2. The Duty-Bearers

Cornwall and Nyamu-Musembi submitted that the Declaration on the Right to Development has created a collective obligation on all states to establish an enabling environment for the realization of the right to development.[7]

The Charter of Economic Rights and Duties of States of 1974 establishes that all states have the responsibility to cooperate in the economic, social, cultural, scientific, and technological fields for the promotion of economic and social progress throughout the world, especially when it comes to developing states.[8] The Charter recognizes the right of every state to benefit from the advances and developments in science and technology for the acceleration of its economic and social development. Therefore, it specifically addresses the issue of transfer of technology in Article 13 by urging all states to promote international scientific and technological cooperation and the transfer of technology. To that end, it exhorts all states to facilitate access by developing states to the achievements of modern science and technology; the transfer of technology; and the creation of indigenous technology

---

[4] Dominique Foray, 'Technology Transfer in the TRIPS Age: The Need for New Types of Partnerships between the Least Developed and Most Advanced Economies' [2009] IPSDS 9.

[5] ibid; Nefissa Chakroun, 'Using Technology Transfer Offices to Foster Technological Development: A Proposal Based on a Combination of Articles 66.2 and 67 of the TRIPS Agreement' (2017) JWIP 2–3.

[6] Watal and Caminero (n 3) 6.

[7] Andrea Cornwall and Celestine Nyamu-Musembi, 'Putting the "Rights-based Approach" to Development into Perspective' (2004) 25 AmUnivLawRev 1415, 1415–37.

[8] The Charter of Economic Rights and Duties of States of 1974—GA Res 3281(XXIX), UN GAOR, 29th Sess Supp No 31, 50.

for the benefit of the developing states in the forms and in accordance with procedures that are suited to their economies and their needs. In that context, Article 13.3 obliges developed states to cooperate with developing states in the establishment, strengthening, and development of their scientific and technological infrastructure and their scientific research and technological activities, so as to help expand and transform the economies of developing states.

The obligation imposed by this provision seems to find concrete operationalization through Article 66.2 of the TRIPS Agreement that demands action from developed states to promote technology transfer to LDCs. Often, the governments of developed states argue that technology is owned by private entities, hence they are unable to compel those entities to transfer technologies to LDCs.[9] Although this assertion may be legitimate, this should not exempt them from pursuing mechanisms to encourage those entities to take action for the benefit of LDCs. International institutions such as the World Bank and the Asian Development Bank are now postulating the concept of a 'capable state' that has the ability to trigger positive dynamics in the private sector, such as complementing private investment whenever necessary and enhancing cooperation and coordination with and within the private sector.[10] With regard to the promotion of technological development, new approaches are advocating for stronger public policies to facilitate, encourage, protect, and induce technological activities in LDCs.[11] This is because new and revisionist or developmental-state approaches claim that market and coordination failures are pervasive in LDCs and that therefore there is a need to redress them.[12]

Furthermore, there are growing calls, including from the UN system and religious communities, for governments to take primary responsibility in promoting the welfare of individuals, their enjoyment of human rights and ensuring the global common good in the world.[13] This is combined with the cosmopolitan ideals that postulate that governments of rich states have a duty to assist other states, especially developing states, in their quest to promote development.[14]

---

[9] European Union, 'Report on the Implementation of Article 66.2 of the TRIPS Agreement' (WTO) IP/C/R/TTI/EU/3 <https://docs.wto.org/dol2fe/Pages/FE_Search/FE_S_S009-DP.aspx?language=E&CatalogueIdList=288475,288461,288378,288403,288245,288236,288039,288037,288041,279272&CurrentCatalogueIdIndex=0&FullTextHash=&HasEnglishRecord=True&HasFrenchRecord=True&HasSpanishRecord=True> accessed 29 October 2023.

[10] Joachim Ahrens, 'Governance and the Implementation of Technology Policy in Less Developed Countries' (2002) 11 EINT 441, 442.

[11] ibid 442–44.

[12] ibid 443.

[13] Brigitte I Hamm, 'A Human Rights Approach to Development' (2001) 23 HRQ 1005, 1016; B Andrew Lustig, 'Natural Law, Property, and Justice: The General Justification of Property in John Locke' (1991) 19 JReligEthics 119, 144; David J O'Brien, 'Sollicitudo Rei Socialis' in Thomas A Shannon and David O'Brien (eds), *Catholic Social Thought: The Documentary Heritage* (5th printing edn, Orbis Books 1992). Vienna Declaration and Programme of Action (adopted 25 June 1993) UN Doc A/CONF.157/23 (hereafter Vienna Declaration).

[14] Simon Derpmann, 'Solidarity and Cosmopolitanism' (2009) 12 EthicalTheoryMoralPract 303; Charles R Beitz, 'Human Rights as a Common Concern' (2001) 95 APSR 269; John D Cameron,

Based on Article 66.2 of TRIPS, it is evident that the governments of developed states have clear and unambiguous responsibilities with regard to the transfer of technologies, and that these ought to be honoured. However, it is worth clarifying that since the technologies to be transferred are owned by private companies, the responsibility of developed states will only take the form of encouragement, promotion, and facilitation.[15] What the governments of developed states are required to do is to create an enabling environment for that process to thrive. The incentives must be effective, entailing that they must be adequate to trigger the dynamics of promoting and encouraging technology transfer to LDCs. In Section B.3 on 'The Nature of Incentives to be Granted', further analysis of the incentives to be provided will be undertaken, and we will also consider how best they can be crafted and implemented to derive the expected results.

3. The Nature of Incentives to be Granted

a) Incentives to companies and institutions in developed states to promote transfer of technologies

The issue of incentives to be granted to companies and institutions in developed states in order to promote the transfer of technology is certainly one of the most debated issues with regard to the implementation of Article 66.2, as evidenced by the heated debates on the issue during the TRIPS Council Sessions and the exchange of documents on the matter between developed states and LDCs.[16] Scholars have proposed that incentives should have the potential to have a positive effect on the transfer of technology. Examples include fiscal incentives; facilitation of access to the market; and capacity-building, funding, and partnerships between companies

---

'Revisiting the Ethical Foundations of Aid and Development Policy from a Cosmopolitan Perspective' in Stephen Brown, Molly den Heyer, and David R Black (eds), *Rethinking Canadian Aid* (2nd edn, University of Ottawa Press 2016) 59.

[15] European Union (n 9).
[16] 'Proposal on the Implementation of Article 66.2 of the Trade-Related Aspects of Intellectual Property Rights (TRIPS) Agreement—Communication from Cambodia on Behalf of the LDC Group' (WTO 2018) IP/C/W/640 <https://docs.wto.org/dol2fe/Pages/FE_Search/FE_S_S009-DP.aspx?language=E&CatalogueIdList=243589,243337,243336,243182,243183,243179,243200,241809,240388,239456&CurrentCatalogueIdIndex=6&FullTextHash=&HasEnglishRecord=True&HasFrenchRecord=True&HasSpanishRecord=True> accessed 12 March 2019; WTO, 'Proposed Format for Reports Submitted by the Developed Country Members under Article 66.2' Doc IP/C/W/561; WTO, 'WTO "Minutes of the Session of the TRIPS Council" (24–25 October and 17 November 2011)' (2012) IP/C/M/67 <https://docs.wto.org/dol2fe/Pages/FE_Search/FE_S_S009-DP.aspx?language=E&CatalogueIdList=42186,38092,99228,99226,99746,108518,103728,96161,98545,96162&CurrentCatalogueIdIndex=0&FullTextSearch=> accessed 12 March 2019; WTO 'Minutes of the Session of the TRIPS Council (27 February 2018)' 25–2 (Doc) (WTO 2018) IP/C/M/88/Add <https://docs.wto.org/dol2fe/Pages/FE_Search/FE_S_S009-DP.aspx?language=E&CatalogueIdList=244469,243545,243337,243336,243182,243183,243179,243200,243245,242032&CurrentCatalogueIdIndex=0&FullTextHash=&HasEnglishRecord=True&HasFrenchRecord=False&HasSpanishRecord=False> accessed 12 March 2019.

and research institutions of both developed and developing states.[17] A closer assessment of the options proposed shows that some could be of immediate relevance, especially those of a fiscal nature. Specific fiscal incentives could include awarding fiscal benefits to firms transferring technologies to developing states; the same tax advantages for research and development performed abroad as for research and development performed at home; ensuring that tax deductions are available for contributions of technology to non-profit entities engaged in international technology transfer; and fiscal incentives to encourage enterprises to employ recent scientific, engineering, and management graduates from developing states.[18]

Scholars have also proposed some non-tax-related incentives, such as establishing special trust funds for the training of scientific and technical personnel and for facilitating the transfer of technologies that are particularly sensitive for the provision of public goods, as well as encouraging research in developing states; allocating public resources to be used to support research into the technology development and technology transfer needs of developing states; and devising grant programmes that offer support to research proposals that meaningfully involve research teams in developing states, presumably in partnership with research groups in donor states.[19] The LDCs also indicated, in October 2011, the 'type of incentives measures for technology transfer' that should have been implemented, such as: financial incentives, grant facilities, loans, tax exemption schemes, investments, technical advice, training, and infrastructure-related incentives.[20] European Union members have identified a further set of incentives that fulfil the requirements of Article 66.2, and a few coincide with the incentives proposed by the scholars and the LDCs. These take the form of promoting projects among private enterprises (such as foreign direct investment, licensing, franchising, sub-contracting, etc); improving access to available information and technologies; supporting common research projects; technology management training to ensure effective incorporation of the transferred technology in a productive capacity; encouraging trade in technological goods; and certain capacity-building initiatives.[21] The pool of incentives proposed seems to be straightforward and very relevant to promoting technology transfer in LDCs.

The World Trade Organization (WTO) Ministerial Conference further directed the TRIPS Council to put in place a mechanism for ensuring the monitoring

---

[17] Keith Maskus, 'Encouraging International Technology Transfer' [2004] IPRSD 1, 33–36.
[18] ibid 35–36.
[19] ibid 33–36.
[20] WTO, 'Proposed Format for Reports Submitted by the Developed Country Members under Article 66.2' (n 16); WTO, 'Minutes of the Session of the TRIPS Council (24–25 October and 17 November 2011)' (n 16).
[21] Foray (n 4) 46.

and full implementation of Article 66.2.[22] The Council then established a reporting mechanism in its Decision of 19 February 2003.[23] Based on the proposal made by the LDCs, the Council also identified the content to be included in the reports, namely the specific legislation involved, the type of incentive, the type of technology and mode of transfer of the technology, the entity making it available, the eligible enterprises, how the system worked in practice, statistics on the use of incentives, the terms of transfer, the recipient LDCs, the extent to which the incentives are specific to LDCs, and information necessary to assess the effects of these measures.[24] Despite this positive development, several issues were flagged regarding the reporting mechanism, such as the irregular submission—or complete failure to submit—of reports and the lack of specific focus on technology transfer directed to LDCs.[25] The reports submitted by developed states to the TRIPS Council do not reveal whether technology received is transferred deliberately as part of the implementation of obligations resulting from Article 66.2 or whether the transfer simply forms part of the routine investment decisions of the business people involved, making it extremely difficult to assess whether developed states are indeed effectively fulfilling the requirements set out in Article 66.2. The LDCs have obviously raised grievances on this issue. Although the situation has improved significantly, there is still a trend towards furnishing long and non-specific reports covering a variety of issues, many of which fall instead under general development aid; hence the establishment of a more effective monitoring system with informational and evaluative functions was recommended.[26] In this context, the LDCs Group requested the Council to take decisions that could force developed states to specify in their reports only the incentives provided to LDC members with regard to technology transfer.[27] Further, the LDCs proposed a specific format for the reports that included sections on 'policy objective and or purpose' and the 'type of incentive measures for technology transfer', as this would facilitate enhanced evaluation processes.

---

[22] WTO, 'Ministerial Conferences—Doha 4th Ministerial Conference—Implementation-Related Issues and Concerns' (2001) WT/MIN/(01)/17 <www.wto.org/english/thewto_e/minist_e/min01_e/mindecl_implementation_e.htm> accessed 19 June 2018.

[23] WTO, 'Implementation of Article 66.2 of the TRIPS Agreement—Decision of the Council for TRIPS' (adopted 19 February 2003) Doc IP/C/28.

[24] WTO, 'Proposed Format for Reports Submitted by the Developed Country Members under Article 66.2' (n 16).

[25] Suerie Moon, 'Does TRIPS Art. 66.2 Encourage Technology Transfer to LDCs?: An Analysis of Country Submissions to the TRIPS Council (1999–2007)' (UNCTAD 2008) Policy Briefing 9 <www.eldis.org/document/A42637> accessed 17 March 2018; Suerie Moon, 'Meaningful Technology Transfer to LDCs: A Proposal for a Monitoring Mechanism for TRIPS Article 66.2' (2011) 9 <https://issuu.com/ictsd/docs/technology-transfer-to-the-ldcs> accessed 1 May 2023; Hans Henrik Lidgard, 'Assessing Reporting Obligations under TRIPS Article 66.2' in Hans Henrik Lidgard and others (eds), *Sustainable Technology Transfer: A Guide to Global Aid & Trade Development* (Kluwer Law International 2012) 43.

[26] Moon, 'Does TRIPS Art. 66.2 Encourage Technology Transfer to LDCs?' (n 25); Moon, 'Meaningful Technology Transfer to LDCs' (n 25); Lidgard (n 25) 43.

[27] 'Proposal on the Implementation of Article 66.2 of the Trade-Related Aspects of Intellectual Property Rights (TRIPS) Agreement' (n 16).

Some of the reports submitted by developed states to the TRIPS Council make vague mention of those incentives. The deeper issue is this: whereas it is unquestionable that some incentives may have been adopted with a view to promoting technology transfer and some initiatives may have been undertaken with that same objective, what remains to be specified is whether those incentives and initiatives have achieved the objective of promoting transfer of technology. It is crucial to address this pressing issue because, apparently, there are no concrete achievements on the promotion or encouragement of technology transfer to an LDC that have been reported on under Article 66.2 and the implementation of this provision is not yet leading to the foreseen objectives that were originally defined by the TRIPS Agreement.[28] Notably, although developed states have been reporting on the initiatives on technology transfer that they have undertaken, and some incentives have been granted in certain areas—such as the environment and water programmes, public health, IP, agriculture and food, energy, and education—in general, there is no feedback from recipient states as to whether these initiatives and incentives have produced any impact on the ground.[29] One state, Rwanda, indicated in its 2010 submission to the TRIPS Council that there is no evidence of its benefiting from specific programmes from developed states.[30] There are accordingly growing fears that the gap between LDCs and the rest of the world might be widening even further.[31] The continuous request of LDCs for extensions of the waiver under Article 66.1 is an unequivocal sign that those states have failed to create a viable technological base as envisaged by Articles 7 and 66.2 of the TRIPS Agreement, and presumably incentives are not sufficiently effective to assist in that regard. Assessing whether incentives are producing any impact is also difficult because there is no linkage between the incentives that purportedly were adopted by developed states and the initiatives that are reported upon to the TRIPS Council. Therefore, doubts may legitimately be cast on whether the initiatives are supported by any incentive at all, and there is no evidence that the private sector may have initiated transactions that involve transfer of technology as a result of the adoption of incentives by any developed state.

To address this challenge, the reporting mechanism would be substantially improved by linking the initiatives purportedly implemented for the benefit of LDCs with specific incentives adopted by governments. Therefore, when reports of developed states are submitted to the Council, they should expressly mention the incentive provided and the initiative that is being reported upon. It would thus be much easier to assess whether the incentive has achieved its results by tracking the impact of the implementation of the initiative on the ground. Formally adopting

---

[28] Chakroun (n 5) 4.
[29] Watal and Caminero (n 3) 15–22.
[30] Rwanda, 'Priority Needs for Technical and Financial Co-operation: Communication from Rwanda' Doc IP/C/W/548.
[31] Chakroun (n 5) 3.

this proposed format would substantially assist the Council in its assessment and deliberations, while simultaneously reducing the amount of work required by LDCs to interpret the reports. Overall, the proposed format would facilitate access and understanding of the reports. This would also exponentially improve the prospects of implementation by developed states and would signal a real commitment to operationalize Article 66.2 as intended.

b) Establishing a mechanism to assess the impact of incentives on transfer of technology

In directing developed states to grant incentives to companies and institutions in developed states, the negotiators of the TRIPS Agreement had in mind that the obligation would stimulate technology transfer to LDCs with the ultimate goal of enabling them to create a sound and viable technological base, as anticipated by Article 66.1.[32] With respect to the interpretation of this provision, some assert that the responsibility of developed states is limited to providing incentives to their own companies while the actual obligation to create a sound and viable technological base remains with the LDCs.[33] To view the obligation in this way would result in an absurdity. It would entail that developed states need not bother themselves with the question of whether the beneficiary LDC is effectively in the process of setting up a sound and viable technological base. It is therefore submitted that such an interpretation is not only absurd, but also misleading, because if it is the TRIPS' overall concern to see that LDCs are able to set up a sound and viable technological base for their development, there must be an interest in assessing whether the support provided by developed states is helping LDCs to achieve that objective. The LDCs expect to see genuine technology transfer and it is thus fundamental that member states refocus their attention to ensuring that the process is not only about fulfilling the obligation of reporting, but is also aimed at accurately measuring the results in terms of promoting and encouraging technology transfer to LDC members. Collaboration is therefore key in this context, because although the responsibility to define the path of development of the state and the assessment of the real impact of the support received through Article 66.2 rests on the shoulders of LDCs, their lack of capacity requires support from developed states.[34]

---

[32] WTO, 'Extension of the Transition Period Under Article 66.1 of the TRIPS Agreement for Least Developed Country Members for Certain Obligations with Respect to Pharmaceutical Products (adopted 6 November 2015)' Doc IP/C/73 <www.wto.org/english/tratop_e/trips_e/art66_1_e.htm> accessed 16 November 2018; 'Declaration on the TRIPS Agreement and Public Health' (adopted 14 November 2001) Doc WT/MIN(01)/DEC/2 <www.wto.org/english/thewto_e/minist_e/min01_e/mindecl_trips_e.htm> accessed 20 November 2019; 'Request for an Extension of the Transitional Period under Article 66.1 of the TRIPS Agreement for Least Developed Country Members with Respect to Pharmaceutical Products and for Waivers from the Obligation of Articles 70.8 and 70.9 of the TRIPS Agreement—Communication from Bangladesh on Behalf of the LDC Group' Doc IP/C/W/605.
[33] Chakroun (n 5) 3.
[34] Watal and Caminero (n 3) 24–25.

One of the examples of initiatives spearheaded by governments of both exporting and importing states that has led to concrete results in terms of transfer of technology to LDCs is the case of the 'Africa Rice Project'.[35] One of the challenges of the African continent is the insufficient production of rice, which is the fastest growing staple food. Despite significant increases in rice production, the continent is still dependent on imports, hence vulnerable to rising prices in global markets. In the bid to support the production of rice in Africa, the government of Japan founded an 'Africa Rice'[36] initiative that involves twenty-seven African states working in partnership. The government of Japan provided funding and the most suitable technology to achieve that objective.[37] The project resulted in increased yields and productivity in African states, benefiting millions of smallholder rice farmers and consumers across Africa, contributing to poverty reduction and food security on the continent. Several technologies related to production, conservation, processing, and commercialization of rice were procured from public and private sectors in Japan and, once transferred to Africa, remained with the beneficiaries, turning the production of rice into a viable and sustainable industry in the beneficiary states.[38] The governments of the recipient states, also realizing that rice production is dominated by small-scale rice millers, encouraged the private sector to invest in efficient rice-processing technologies, such as 'mini-rice mills' with built-in capacity for de-stoning, polishing, and sorting homogeneous high-quality rice to stimulate the local rice value chain and improve quality.[39] To that end, those governments provided incentives for processors to upgrade their technologies, such as duty-free imports on processing equipment, tax concessions, or access to finance.[40] Import tax reductions were also provided on the machinery that improved labour efficiency and on fertilizers.[41] This case study does not aim to exhaustively describe the 'Africa Rice' initiative and the targeted support by one of the donors, the government of Japan. Rather, the elements emphasized lead to the conclusion that the success of the initiative is the result of an integrated approach to technology transfer and targeted and coordinated incentives from both the exporting and importing states. Decisive in this initiative was the clear identification of needs that were common to the entire continent, related in this specific case to insufficient supply of rice to the ever-growing population. The government of

---

[35] WTO, 'Minutes of the Council for TRIPS—Communication by the Delegation of Benin (27–28 February 2018)' Doc IP/C/M/89/Add.1, 26; 'About Us—CARI: Competitive African Rice Initiative' <www.cari-project.org/> accessed 29 April 2021.

[36] Matty Demont and others, 'From WARDA to Africa Rice: An Overview of Rice Research for Development Activities Conducted in Partnership in Africa' in Eric Tollens and others (eds), *Realizing Africa's Rice Promise* (CABI Publishing 2013).

[37] 'About Us—CARI: Competitive African Rice Initiative' (n 35).

[38] WTO, 'Minutes of the Council for TRIPS—Communication by the Delegation of Benin' (n 35).

[39] M Wopereis and others, 'Realizing Africa's Rice Promise: Priorities for Action' in M Wopereis and others (eds), *Realizing Africa's Rice Promise* (CABI Publishing 2013) 430.

[40] ibid.

[41] ibid.

Japan mobilized technologies from a wide spectrum of providers in both the private and public sectors through the adoption of appropriate incentives. Recipient states also contributed by identifying specific needs in terms of modern technologies, especially in rice processing, and provided their own incentives to facilitate importation and use of the technologies. These actions were further blended with consistent agriculture and trade policies, including price protection for local rice; improved capacity at research, extension, processing, and marketing levels; testing and transferring technologies in a targeted and systematic manner; and identification or development of appropriate technology options that address end-users' needs, accompanied by adaptive research and development to respond to technologies needed locally.[42]

The 'Africa Rice' model should be considered in other sectors in order to develop an efficient mechanism of transfer of technologies that has an everlasting impact and is sustainable. Therefore, inspired by the successful experience of 'Africa Rice', a mechanism is proposed that can efficiently promote the transfer of technology for the benefit of LDCs. This may be achieved through the establishment of 'Clusters of technology transfer' as a smart and integrated approach to guiding governments in channelling their incentives to a mechanism that can effectively attract technologies to the entire continent, regions, or nations of Africa.

c) The proposal for the establishment of clusters of technology transfer

The patterns of technological need are similar in the various states on the African continent and across different regions. Therefore, efforts to promote technology should be designed taking into account those similarities and developing clusters based on geographic proximity as well as sectoral specialization.[43] Drawing on the successful experience of 'Africa Rice', clusters aimed at addressing food insecurity[44] and the growing population[45] on the African continent should be developed. Fighting food insecurity requires the development not only of rice production but also of other staple foods (cereals, roots, and tubers) that are insufficient. Therefore,

---

[42] Demont and others (n 36) 11.
[43] Douglas Zhihua Zeng (ed), *Knowledge, Technology, and Cluster-Based Growth in Africa* (World Bank Publications 2008); Dorothy McCormick, 'Industrialization Through Cluster Upgrading: Theoretical Perspectives' in Banji Oyelaran-Oyeyinka and Dorothy McCormick (eds), *Industrial Clusters and Innovation Systems in Africa: Institutions, Markets, and Policy* (United Nations University Press 2007) 20; Eva Galvez, 'Agro-Based Clusters in Developing Countries: Staying Competitive in a Globalized Economy' (2010) 51–60 <www.fao.org/sustainable-food-value-chains/libr ary/details/ru/c/267092/> accessed 18 November 2018; Hubert Schmitz, 'On the Clustering of Small Firms' (1992) 23 IDS Bulletin 64, 64–69.
[44] The Global Hunger Index 2018 reveals that in Sub-Saharan Africa, the levels of undernourishment, child stunting, child wasting, and child mortality are unacceptably high, see Klaus von Grebmer and others, 'Global Hunger Index 2018: Forced Migration and Hunger' (2018) 12 <https://developme nteducation.ie/resource/global-hunger-index-2018-forced-migration-and-hunger/> accessed 17 January 2019.
[45] 'World Population Prospects—Population Division—United Nations' <https://population. un.org/wpp/> accessed 9 May 2023.

it makes sense that apart from the 'rice cluster' other staple food clusters be developed to ensure food security and that the necessary technologies be mobilized through transfer of technology. Indeed, the agricultural sector throughout Africa is characterized by low technology and non-mechanized farming, which negatively impacts output and productivity.[46] Some of the clusters that could be established to promote transfer of technology in the staple foods sector include 'maize cluster', 'millet cluster', 'cassava cluster', 'potato cluster', 'tomato cluster', 'yam cluster', and 'groundnut cluster'.[47] Clusters can also be established for the production and commercialization of export crops such as cocoa, coffee, tea, cotton, flowers, and fruits. Each cluster may require a specific technology for the specific phases of the value chain, such as tilling, sowing, harvesting, processing, conservation, storage, packaging, and transportation. The cluster could be limited to one state or could comprise neighbouring states, a region, or the entire continent as evidenced by the case of 'Africa Rice' that included twenty-seven states from almost all regions of Africa. Technologies to develop these clusters are readily available in emerging states and in the developed world.

The advantage of establishing a 'cluster of technology transfer' for a specific sector is to make it easier to identify all the technologies necessary in the entire value chain; determine those states in need; and ascertain possible technology providers, states of origin, necessary quantities, and costs. Certainly, in this way, economies of scale could make the continental, regional, or national cluster more attractive for investors and donors from the developed world. Furthermore, once the cluster is identified, both exporting and importing states could take informed decisions with regard to possible incentives to be provided and other ancillary policies to be adopted, including research and capacity-building to facilitate absorption and adaptation and possible spillovers in the market of the technologies, thus making the cluster sustainable in the long run.

Technology clusters may be established in other economic sectors based on the priorities of development of each country. Some sectors in which Africa may have a comparative advantage and that can benefit from 'technology transfer clusters' include: 'wood processing cluster', 'mining processing cluster', 'fishing cluster', 'textile cluster', and 'handcraft cluster'. It is also submitted that a technology transfer cluster may be established to generate start-ups in emerging areas such as information and communication technologies, the 'Internet of Things', artificial intelligence, cloud computing, and big data, thus promoting the development of new sectoral activities on the continent.

---

[46] Jessica Fanzo, 'The Nutrition Challenge in Sub-Saharan Africa: United Nations Development Programme' (2012) Working Paper No 32 <www.undp.org/africa/publications/nutrition-challenge-sub-saharan-africa> accessed 2 January 2019.

[47] The suggestion of these clusters is based on high demand for those staple foods, as highlighted in the 'Staple Foods Production in Africa' graph developed by Fanzo, ibid 35.

4. The Use of Information and Communication Technologies

To further improve access to information provided by developed states on the issue, the use of information and communication technologies is under consideration at the WTO. Accordingly, the idea of establishing a digital tool is being discussed at the WTO TRIPS Council, and contributions on its format and content are being sought.[48] This book provides its own contribution to this topical issue. Accordingly, it is submitted that the proposed online platform should not be limited to a repository of the reports presented by developed states in digital format. The online platform should be structured as a database that would systematize the information and facilitate its access and utilization. The database should therefore allow the information to be retrieved based on different parameters. Additionally, the database should become a match-making tool for the priority areas defined by the LDCs, specific technological needs to meet those national priority development areas, incentives provided by the developed states to stimulate interest in making available the necessary technologies to match the needs of LDCs, project proposals that match those needs, and technical assistance and funding opportunities for technology transfer. A database format would also facilitate the identification of best practices that could be emulated by other stakeholders. Viewed in this way, the database would become a 'meeting point' between the needs and aspirations of LDCs and potential well-wishers that possess the technology needed.

Another challenge that needs to be resolved is the fact that, for a long time, developed states have been subject to enormous pressure to segregate technology transfer that is provided exclusively to LDCs from initiatives benefiting other developing states. This is because it has emerged that developing states are benefiting more than LDCs from technology inflows. Available data shows, for example, that in 2012, out of the 245 reported programmes, only forty-seven had been specifically designed for LDCs, corresponding to less than 19 per cent, and even fewer were related to technology transfer. By 2013, LDCs again reported that out of 145 programmes or projects presented in the reports only 10–15 per cent were related to technology transfer.[49] Apparently, developing states are charting a virtuous path and are in a better position to attract technology inflows than LDCs as a result of the improvement of their absorptive capacities.[50] This in turn is facilitating the dissemination of technologies and spillovers within their jurisdictions. Contrasted against that, LDCs are trapped in a vicious cycle of low levels of foreign direct investment and trade and poor quality of their absorptive capacities, hence they are unlikely to attract any significant amount of foreign technologies.[51] It has therefore

---

[48] WTO, 'Minutes of the Session of TRIPs Council (10–11 October 2013)' Doc/IP/C/M/74/Add.1; 'Trade-Related Aspects of Intellectual Property Rights—Welcome to the e-TRIPS Gateway' <https://e-trips.wto.org/> accessed 9 May 2023.
[49] WTO, 'Minutes of the Session of TRIPs Council (10–11 October 2013)' (n 48) 17.
[50] Foray (n 4) 4–6, 42–43.
[51] ibid 7, 42–43.

been established that states without an enabling environment for investment and with poor absorptive capacities have fewer chances of stimulating inward technology flows.[52]

Developed states have argued that segregating the initiatives undertaken in technology transfer was not workable. For example, the European Union (EU) has been insisting that, because it favours regional integration in its operations, it usually targets groups of states or regions. In light of this, the EU affirms that it does not have single LDC-specific technology transfer programmes.[53] The main submission is that a database would assist member states in overcoming this challenge, in that its format would allow users to apply filters in order to identify or isolate beneficiary states in which they are particularly interested. Consequently, targeted programmes on technology transfer could be developed to specifically benefit the states in need.

C. Institutional Arrangements to Support the Implementation of Article 66.2

1. National Institutions to Monitor the Implementation of Article 66.2

One of the difficulties that developed states are encountering in fulfilling the obligations related to the reporting mechanism established by Article 66.2 is in relation to gathering information from the various government agencies and compiling it as a consolidated document for further submission to the TRIPS Council. To date, the reports submitted to the TRIPS Council by developed states number more than 5,000 pages and in every new session 200 additional pages are submitted.[54] That compounds the difficulties that the LDCs experience in respect of analysing such an enormous amount of information and making proper sense of it.

This means that there is a need to set up mechanisms both upstream and downstream to facilitate the preparation of the reports by developed states and also improve consumption of information by relevant institutions in LDCs. This mechanism is not available currently, hence LDCs have expressed their dissatisfaction with the reports as they appear more like a mere compilation of information coming from different agencies and thus lack coherence.[55] In view of this, LDCs struggle to read and comprehend them. This is partly because the reports include

---

[52] Maskus (n 17) 7.
[53] WTO, 'Minutes of the Session of the TRIPS Council (8–9 November 2016)' Doc IP/C/M/83/Add.1, 38; WTO, 'Minutes of the Session of the TRIPS Council (27 February 2018)' Doc IP/C/M/88/Add.1, 22.
[54] Watal and Caminero (n 3) 13; WTO, 'Minutes of the Session of the TRIPS Council (27 February 2018)' (n 53) 22; WTO, 'Minutes of the Session of the TRIPS Council (8–9 November 2016)' (n 53) 38.
[55] WTO, 'Minutes of the Session of the TRIPS Council (27 February 2018)' (n 53) 20–27.

activities of technical assistance under Article 67, and thus do not reveal much in terms of a deliberate effort to promote systematic or sustainable technology transfer to LDCs.[56]

It was submitted that the absence of a clear institutional arrangement upstream which can gather, analyse, assemble, and systematize data on technology transferred to LDCs with the ultimate objective of drafting coherent reports to be submitted to the TRIPS Council has a significant impact on the quality of the documents produced. An efficient mechanism is the establishment of relevant focal points on transfer of technologies in developed states to coordinate structured support to LDCs. This may be in the form of helpdesks in the existing international development agencies and in departments of industry, international cooperation agencies, or the bodies responsible of science and technology and innovation.

Downstream, the lack of capacity to analyse the reports submitted and to assess their effectiveness in promoting transfer of technology to LDCs will also perpetuate suspicions that these documents are not useful. Therefore, with respect to the purpose of the present book, the exercise of ascertaining whether developed states are implementing the obligation prescribed by Article 66.2 cannot depend only on the quality of the documents produced, but rather more on the ability to analyse them and determine whether transfer of technology is effectively occurring on the ground. Accordingly, individual LDCs should strive to set up national institutional frameworks to monitor the impact of the support that they are receiving from developed states that claim to facilitate technology transfer. For example, the TRIPS Agreement has recommended that WTO members set up contact points on IP at the national level and advise the TRIPS Council of these contact points.[57] It was therefore recommended that those contact points work hand in hand with the institutions responsible for technology transfer and innovation in their respective states to track the initiatives that are highlighted in the reports submitted by developed states to the TRIPS Council, with a view to assessing their impact in promoting technology transfer.

One case is illustrative of concrete action that may be undertaken in this regard: the assessment of the period of fourteen years from 2003 to 2016 reveals that of the 221 reports submitted by developed states to the TRIPS Council under Article 66.2, Mozambique appears as a recipient of support 135 times, a higher number than any other LDC.[58] Yet, there is no evidence that Mozambique has actually been the major beneficiary of technology transfer initiatives and as a result is on the right track on technological catch-up and innovation. Therefore, this phenomenon must be assessed so that accurate conclusions can be drawn on

---

[56] ibid 25.
[57] WTO, 'Intellectual Property (TRIPS)—Notifications: Contact Points' <www.wto.org/english/tratop_e/trips_e/trips_notif5_art69_e.htm> accessed 29 April 2021.
[58] Watal and Caminero (n 3) 21.

the meaning of the assistance provided and its significance. It is therefore recommended that the Ministry of Science and Technology, which is responsible for issues related to technology transfer and innovation in Mozambique, liaise with the IP contact office, the Industrial Property Institute, in order to gather the information emanating from the TRIPS Council and undertake a thorough analysis of the reports over the specified period, to determine whether technology transfer has effectively occurred and is having a positive impact towards creating a sound and viable technological base in the country.

2. Impact-Assessment Studies on the Implementation of Article 66.2

Establishing mechanisms to systematically assist LDCs in measuring the impact of the initiatives reported by the developed states to the TRIPS Council is crucial in order to guide future developments. This must be further complemented by evidence that a policy objective of a specific state related to development was attained and a specific local technical problem solved. That can be achieved not only by linking incentives to initiatives in the reports but also by emphasizing the unequivocal benefit that has accrued to an LDC as a result of the technology transferred, as shown in the case of 'Africa Rice'. As such, empirical studies must be undertaken to assess the impact of the implementation of the projects that were included in the developed states' reports to the TRIPS Council.[59] The book maintains that this kind of assessment is not only necessary, but urgent, as it could become a useful tool to showcase progress (if any) in technology transfer in addition to assisting in determining whether the technology transferred is having the much-desired impact of creating a sound and viable technology base. Therefore, the studies should assess concrete projects implemented in some LDCs and ensure that these are effectively facilitating access to technologies and that technological capabilities are being developed in the targeted states. This could inform future projects in the state and would ensure that the most successful and relevant projects are replicated in other states.

These studies would also be useful for the developed states if they clearly highlight the types of incentives that were adopted and the manner in which they effectively contributed to promoting transfer of technologies to LDCs. Such incentives and successful implementation could be used as best practices to be emulated by other states and would drive future efforts to adopt new incentives and initiatives that would definitely facilitate the implementation of Article 66.2. Finally, a thorough assessment of the current dynamics of technology transfer may shape the

[59] ibid 24–25.

future course of negotiations with regard to the best way of implementing Article 66.2 and may even assist in recommending future improvements to the provision.

The relevance of assessments at field level is well established and ought to be considered. The burning question, however, is how the studies should be conducted and by whom? It is submitted that since this is a crucial issue for national development, it deserves to be prioritized. Therefore, individual LDCs should strive to set up their national institutional frameworks or better coordinate the existing ones to assess the impact of the support that they are receiving from developed states which claim to facilitate technology transfer. In order to track the initiatives that are highlighted in the reports submitted by developed states to the TRIPS Council, national focal points established under Article 69 should collaborate closely with the institutions in charge of technology transfer and innovation in their respective states. This will allow them to assess the impact of initiatives to promoting technology transfer into their respective statesThe reports of developed states on the implementation of Article 66.2 are readily available on the WTO website.[60] The designated entities from LDCs should collect information regarding the support that was provided to their respective states; locate the projects implemented in the field; and assess their status, focusing their attention on the transfer of technology component and its impact on supporting the state to create a viable technological base. In Box 5.1, guidelines are proposed for field impact assessment that should be conducted in LDCs with a view to assessing the benefits derived from the implementation of Article 66.2 of TRIPS at national level.

Least Developed Countries may claim that they lack capacity to undertake those comprehensive studies; thus, an alternative option is to request technical assistance under Article 67. Indeed, Article 67 on technical assistance is a crucial enabler for the implementation of Article 66.2 and cannot be dissociated from it. LDCs shall take advantage of the available opportunities to formally request technical assistance to undertake impact assessment studies of the projects included in the reports submitted by developed states to the TRIPS Council in fulfilment of the obligation emerging from the implementation of Article 66.2.

In 2005, the TRIPS Council called on LDCs to identify their priority needs to facilitate technical and financial cooperation in order to assist them in taking necessary steps to implement the TRIPS Agreement.[61] Out of the forty-seven LDCs that are members of the WTO, only Sierra Leone and Uganda submitted their priority list by the prescribed deadline of 1 January 2008. Another seven states (Bangladesh, Madagascar, Mali, Rwanda, Senegal, Tanzania, and Togo) submitted

---

[60] 'By Topic' <https://docs.wto.org/dol2fe/Pages/FE_Browse/FE_B_009.aspx?TopLevel=4482#/> accessed 9 May 2023.
[61] 'Decision of the TRIPs Council for Extension of the Transition Period under Article 66.1 for Least-Developed Country Members' (adopted 29 November 2005) Doc IP/C/40.

Box 5.1 Guidelines for the Field Impact-Assessment of the Technologies Transferred in the Context of the Implementation of Article 66.2 of TRIPS

The WTO Ministerial Conference meeting at its 4th Session in November 2001 reaffirmed that the TRIPS Agreement is mandatory and directed the TRIPS Council to put in place a mechanism for ensuring the monitoring and full implementation of Article 66.2. Following that directive, in 2003 the TRIPS Council established a reporting mechanism on the practical functioning of the incentives provided to companies for the transfer of technology.

The Reports presented every three years, supplemented by updates every year, provide valuable information on the initiatives undertaken by developed states to promote technology transfer to LDCs. However, the impact of these initiatives in individual LDCs is as yet unknown due to the lack of feedback from the beneficiary LDCs.

These guidelines aim to assist beneficiary LDCs in assessing the impact of initiatives in their respective territories. LDCs wishing to undertake field assessment studies shall consider observing the following:

1. Objective of the field assessment
The field assessment aims at identifying the initiatives or projects undertaken by private companies or public institutions from developed states to transfer technologies to the LDC as a result of the implementation of Article 66.2 of TRIPS and determine their concrete impact at national level. The field assessment will further determine whether the technology transfer transaction was supported by concrete mechanisms to enable assimilation, absorption, and adaptation by and to the local needs and generate spillovers in the local context that may facilitate further innovation.

2. Target of the field assessment
The assessment shall target all initiatives or projects that were included in the reports of the developed states to the TRIPS Council that purportedly benefited the LDCs undertaking the field assessment.

3. Periodicity
Overall field assessment shall be undertaken every three years after the presentation of new reports at the TRIPS Council by developed states. Annual assessments shall also be conducted to analyse the updated information presented by developed states during the annual reviews.

4. Institutional arrangements to undertake the field assessment

Governments of the LDCs shall establish focal points in government institutions responsible for innovation and science and technology to monitor the initiatives that benefit their respective states as a result of the implementation of Article 66.2 of TRIPS. In conducting the field assessments such institutions may co-opt any other relevant public institutions, research and development and academic entities, and the private sector interested in the technological areas that are the subject of technology transfer.

5. Format and content of the output

The field assessment shall produce a report that identifies the initiatives undertaken that directly benefited the state and shall include, among others: the state-provider; the incentive established by the developed state that prompted the transfer of technology initiative; the company or public entity undertaking the initiative as a result of the incentive; the beneficiary company or entity in the LDC; the technological sector covered; the value of the technologies transferred; the modalities of transfer; whether the technology was transferred with or without cost to the beneficiary, and the amount involved; the mechanisms put in place to ensure that the technology transferred is assimilated, adapted, and promotes innovation in the beneficiary state; the impact of the technology transferred in solving state technological challenges; its contribution to the overall technological development of the state; whether the technology transferred fits into technological development priorities of the state; whether the technology transfer process was successful or unsuccessful; the challenges that the state faced in receiving and assimilating the technology; the monitoring mechanism that was set up to ensure that the transfer of technology process is sustainable and long-lasting.

6. Use of the field assessment outputs

The results of the field assessment shall be submitted to the ministries responsible for innovation and science and technology and the liaison ministry with the TRIPS Council for further action at the national, international, and multilateral level.

a) At the international and multilateral level

The government of the LDC should consider presenting its own position with regard to the impact of the implementation of Article 66.2 in the relevant forums, including but not limited to the TRIPS Council; the workshops organized by the WTO on technology transfer; at the Ministerial Conference; and any other bilateral forums related to trade, development, and technology transfer. It is expected that the LDC that has undertaken the field assessment will present its recommendations on how the incentives and the initiatives resulting from the implementation of Article 66.2 may be improved to produce better results in its own state.

b) At the national level

The results of the field assessment reports shall also be used by the government to sensitize national stakeholders on the existing technologies and on the need to exploit them for the benefit of the national industry and for academic and research purposes. The results should also inform the adoption of national incentives to attract technologies from developed states.

their list during the period 2010–13. No information whatsoever was submitted by the rest of the LDC members.[62]

Of the eight African states that presented their priority needs to the TRIPS Council, Sierra Leone appears to have submitted a straightforward request for a targeted assessment of innovation capabilities to be undertaken in the following terms:

> To support the creation of a sound and viable technological base in Sierra Leone, a scoping study should be undertaken to examine how domestic creativity, innovation and transfer of technology can best be stimulated through reinforcement of domestic policies, incentives, private sector associations, and capacity building programmes, including the IPR system, and through more targeted measures taken by the developed countries in line with their obligations under Article 66.2 of the TRIPS Agreement.[63]

The Sierra Leonean government worked in close cooperation with ICTSD-Saana Consulting and the United Kingdom Department for International Development (DFID) and produced a report that revealed a very low technological base, institutional weakness, a pressing need for human social and economic development, and limitations at all levels of the education system as its major problems.[64] The

---

[62] See list of states that submitted a priority list of needs at <www.wto.org/english/tratop_e/trips_e/ldc_e.htm> accessed 16 May 2018.

[63] 'Priority Needs for Technical and Financial Cooperation—Communication from Sierra Leone' Doc IP/C/W/499.

[64] Ahmed Bashir, 'An Analysis of the Existing Mechanism of Transfer of Technology from the Developed to the Developing Countries underthe TRIPS Agreement' (Queen Mary University of London 2011) 7 <www.academia.edu/856420/ International_Transfer_of_Technology_under_the_TRIPS_Agreement?auto = download> accessed 9 September 2019.

report recommended that a study be undertaken to examine how domestic creativity, innovation, and transfer of technology could be stimulated through domestic measures and external help from developed states under Article 66.2 of TRIPS; that technical and financial assistance be provided to design, implement, and evaluate systems of education and conduct campaigns for raising awareness in IP management for small to medium enterprises and using IP for development; the that a Patent Information System (PIS) be developed to support innovation and technology transfer through the identification of relevant technologies for Sierra Leone's key sectors such as mining, fishing, forestry, and agriculture; and that financial and technical help be provided for human resource capacity-building and building infrastructure of an administrative and judicial nature for IP rights protection.[65]

Rwanda also requested assistance to carry out a survey that would provide a clear baseline for understanding the current levels of innovation, sources of funding, and incentives.[66] A study was also conducted by ICTSD-Sanaa Consulting in 2007 which provided very insightful recommendations with regard to the relevant actions to stimulate technology transfer and technological development, namely: helping firms to identify relevant technologies from patent information and to identify protectable subject matter and address issues relating to licensing; providing a patent information service in relation to patents in Rwanda and internationally; assisting Rwandan industry in identifying relevant public domain technologies; setting up IP management and international technology transfer help desks; and for supporting patenting and dissemination of patented technical information.

Similarly, Senegal requested a study to examine domestic measures, such as tax incentives to promote innovation and opportunities for technology licensing and contract-based research, development of local incentives for foreign direct investment, and design of support programmes in priority research and development sectors.[67]

In light of the reports produced that were shared at the TRIPS Council, it would have been expected that the needs presented would have shaped the initiatives of support by the developed states presented in subsequent submissions. As it stands, there is not yet evidence of any projects initiated on the basis of the reports of the LDCs. Developed states have maintained their support based on their own programmes of technical assistance for overseas development, without any regard for the needs of the beneficiary states.[68]

---

[65] ibid.
[66] ibid; Rwanda, 'Priority Needs for Technical and Financial Co-operation: Communication from Rwanda' (n 30).
[67] Rwanda, 'Priority Needs for Technical and Financial Co-operation: Communication from Rwanda' (n 30).
[68] Bashir (n 64) 12–13.

In this context, it appears to be appropriate to recommend that future initiatives related to the implementation of Article 66.2 be guided by the concrete needs that are presented by LDCs in the TRIPS Council. Furthermore, the support should cover all LDCs, and therefore it is essential that these states are encouraged to submit reports. Moreover, this process of continuous submission of reports must be prioritized, given that needs and priorities can change over time. The reports also reveal that the needs submitted were broad. Accordingly, it is recommended that a specific request regarding impact assessment studies of the projects related to technology transfer as a result of the implementation of Article 66.2 be made.

An additional potential source of support is the newly created Technology Bank for LDCs created by the UN General Assembly to facilitate access to appropriate technologies for LDCs.[69] This book proposes that the potential of the Technology Bank be fully harnessed to support the implementation of Article 66.2. For example, the Technology Bank has defined as its focus for the initial period to undertake baseline reviews and technology needs assessments on science, technology, and innovation in LDCs. The Technology Bank could therefore direct its efforts towards the assessment of the impact of the implementation of Article 66.2 at the national level and advise LDCs on those matters.

In conclusion, the main focus of the current debates on the implementation of Article 66.2 is on the effectiveness of incentives adopted by developed states to assist LDCs to build a sound and viable technological base for their development. For a long time, LDCs did not achieve much in the negotiations pertaining to the issues highlighted. Even more worrying is the fact that they did not obtain the much-desired technology transfer either. With the adoption of the recommendation provided in this book, it is expected that some improvements will occur in the implementation of Article 66.2, and LDCs will witness more inflows of technology from the developed world.

3. The Proposal for a WTO Advisory Centre for Technology Transfer and Innovation

Ultimately, it is argued that the Ministries of Science and Technology in the LDCs may not have the necessary skills to undertake the assessment of the technology flows into their respective territories as suggested in Section C.2 on the 'Impact-Assessment Studies on the Implementation of Article 66.2'. One of the solutions to this challenge consists of the establishment of a WTO Advisory Centre for

---

[69] United Nations, 'Resolution 71/251 on the Establishment of the Technology Bank for the Least Developed Countries'.

Technology Transfer and Innovation (ACTTI) to assist LDCs to take full advantage of the flows of technology from developed states so as to build a technology base for their development. The idea of an Advisory Centre for Technology Transfer mimics two previous experiences in the international arena. The first was the Advisory Service on Transfer and Development of Technology (ASTT) established by the UNCTAD's Transfer of Technology Division in 1976 to provide advice, technical assistance, and operational assistance to developing states on the transfer and development of technology.[70] The second is the existing Dispute Settlement Mechanism Advisory Centre on WTO Law (ACWL). It should be recalled that ACWL was established when member states realized that LDCs and developing states were not in position to use the DSM due to the high costs involved. The ACWL acts as a law firm at a concessional fee and provides legal advice to developing states in the dispute settlement process.[71] Taking advantage of this facility, many developing states, including India, Thailand, Ecuador, Guatemala, and Indonesia, have used the services to initiate disputes before the DSM of the WTO.[72] Similarly, the proposed ACTTI intends to assist LDCs to fully benefit from the provisions of Articles 7 and 66.2 of TRIPS.

The proposal of an ACTTI is limited to advisory services with a view to facilitating access to technologies by LDCs whenever such an opportunity emerges. For the purposes of this book, the main objective would be linked to the full implementation of Article 66.2 by assisting LDCs to assess the impact of the incentives provided by developed states to effectively transfer technologies to LDCs. However, ACTTI could undertake any other initiative that would assist LDCs in their quest to obtain access to technologies. This seems to be achievable in scope and within the capacity of the WTO.

A less ambitious version of this proposal would consist of expanding the mandate of the ACWL beyond the DSM. Apparently, ACWL is mandated to provide opinions on any legal issues arising in WTO decision-making and negotiations. Therefore, technology transfer could also fall under the competence of this institution. The findings of this book suggest that it would be less attractive to advocate for the use of this facility because undertaking impact assessment studies on technology transfer goes beyond the realm of providing a legal opinion on the implementation of Article 66.2. A specialized institution on technology matters appears to be more appropriate to achieve the intended results, hence the preference for establishing ACTTI.

---

[70] UNCTAD, *Handbook on the Acquisition of Technology by Developing Countries* (United Nations 1978) 57.
[71] Agreement Establishing the Advisory Centre on WTO Law, Annex II (signed 13 November 1999, entered into force 15 July 2001) WT/GC/W/446.
[72] Atul Kaushik, 'Dispute Settlement System at the World Trade Organisation' (2008) 43 EPW 26, 27.

D. Enforcement of Article 66.2 through the WTO Dispute Settlement Mechanism

1. The WTO Dispute Settlement Mechanism and TRIPS

One of the unique features of the WTO system is the Dispute Settlement Mechanism.[73] The DSM is a mandatory and unified dispute settlement system, which applies to all WTO Agreements, including TRIPS. Therefore, any violation of a mandatory obligation under the WTO Agreements by any WTO member may trigger a complaint by another member.[74] Appendix 1 of the Understanding on Rules and Procedures Governing the Settlement of Disputes (the Understanding) includes, among the agreements that are covered, the TRIPS Agreement. Therefore, based on Article 23 of the Understanding, a member state can lodge a complaint if any benefit accruing to it directly or indirectly is being nullified or impaired or the attainment of any objective of the Agreement is being impeded as the result of the failure of another contracting party to carry out its obligations.

The implementation of the Understanding potentially confer on LDCs, individually or collectively, the right to legitimately raise a complaint to the WTO Dispute Settlement Body arguing that developed states are violating the obligation placed on them under Article 66.2 of TRIPS to provide incentives to companies so as to promote transfer of technologies to LDCs. A hypothetical scenario sheds more light on the LDCs' quest and possible decisions by the WTO.

LDC X is a well-known exporter of ceramics and unique pottery products that has a competitive advantage in international markets but is still grappling with a technological challenge that could enhance productivity and economies of scale. The challenges of the ceramics and pottery industry of state X are further exacerbated by the recent introduction of environmentally friendly practice requirements for products to enter region Z in the developed states. Company A from developed state Y has already found the technology that solves the technological obstacle hampering mass pottery production by state X and to fulfil the environmentally friendly practice requirements imposed by region Z. However, the technology benefits from intellectual property protection for the next fifteen years, and

---

[73] Monika Bütler and Heinz Hauser, 'The WTO Dispute Settlement System: A First Assessment from an Economic Perspective' (2000) 16 J LawEconOrgan 503, 504–505; Peter Van den Bossche, 'The Doha Development Round Negotiations on the Dispute Settlement Understanding' (WTO 2003).

[74] Xavier Groussot and Thu-Lang Tran-Wasescha, 'TRIPS Article 66(2)—Between Hard Law and Soft Law?' in Hans Henrik Lidgard, Jeffrey Atik, and Tu Thanh Nguyen (eds), *Sustainable Technology Transfer: A Guide to Global Aid & Trade Development* (Kluwer Law International 2012) 19; Bütler and Hauser (n 73) 509; Malebakeng Agnes Forere, *The Relationship of WTO Law and Regional Trade Agreements in Dispute Settlement: From Fragmentation to Coherence* (Wolters Kluwer: Kluwer Law International 2015) 128–29.

company A is refusing to transfer the technology without payment of the royalties based on the market value of the technology and subject to further annual royalties. The market value of the technology and the royalties demanded by company A are prohibitive for state X. All private negotiations have failed to yield any tangible results. The government of state X has also used diplomatic channels to sensitize the government of state Y so that it will convince company A to sell the technology at a discounted price to enable state X to enter the international market and compete equally with other companies. The government of state Y was cooperative but lamented the fact that the technology is owned by company A and nothing can be done to force that company to transfer the technology to state X. No further action was taken by the government of state Y to persuade company A to share the technology with state X. The government of state X is now considering taking advantage of Article 66.2 of TRIPS that requires governments of developed states to act in such situations by providing incentives to enterprises and institutions in their territories that may encourage those entities to share the technologies with LDC members in order to enable them to create a sound and viable technological base. The government of state X indeed considers that possessing such technology could be a game-changer because this would create a sound technological base for the development of a high-earning product that could transform the state's economy. Is this a plausible and strong cause of action?

There appears to be no doubt that this is a typical situation that many LDC members face in their economies. An upgrade in some of their obsolete industries could catapult them into a new dimension that could change their economies significantly and enable them to compete with other states. In this case, the complainant LDC member X would sustain its cause by alleging that another member of WTO, (a developed state) Y, is not carrying out its obligations as specified by TRIPS and as a result of that, the expected benefits are being nullified or impaired, or an objective of TRIPS is being impeded.[75] This can be based on Article 23 of the Understanding in that one of the objectives of the Agreement is to promote and encourage technology transfer to LDC members in order to enable them to create a sound and viable technological base. This objective is nullified or impeded as a result of the behaviour of developed state Y. This could amount to a material violation of the obligation imposed by Article 66.2 on developed states; hence, LDC member state's plea could be judged to be legitimate and well grounded.

---

[75] Antony Taubman, *A Practical Guide to Working with TRIPS* (Oxford University Press 2011) 137; Andrew C Michaels, 'International Technology Transfer and TRIPS Article 66.2: Can Global Administrative Law Help Least-Developed Countries Get What They Bargained For?' (2009) 41 GJIL 252.

Despite the existence of such a potential avenue and a strong cause of action, no African member of the WTO has ever initiated any procedure under the DSM, including complaints related to non-compliance with Article 66.2 on technology transfer.[76] The hypothetical scenario provided shows that there is indeed a plausible cause of action that could be pursued by the LDC and success in this case at the Dispute Settlement Body is foreseeable. In the circumstances, it is pertinent to discuss why it is that African states have not used the DSM. Moreover, some recommendations to overcome those challenges are also adduced.

2. Challenges in Exploiting the Potential of the Dispute Settlement Mechanism to Enforce the Claims of LDCs

The LDC group has been complaining for years that its member states are facing challenges in using the DSM due to its complexity and the excessive cost of litigation.[77] Some scholars raise doubts about the legitimacy of this complaint as it appears to be more the lack of political will to pursue cases through the system than the lack of capacity or knowledge to lodge complaints.[78] In relation to the concerns regarding complexity of processes, at the TRIPS Council Session in February 2018, the LDCs' group amplified the concern by raising the ambiguity of concepts and the consequent difficulties associated with

sustaining an action before the WTO Dispute Settlement Body.[79] These objections were refuted and it was established that the language used in the TRIPS provisions unequivocally imposes obligations with legally binding force on its member states.[80] The mandatory nature of Article 66.2 can also be seen in paragraph 11.2 of

---

[76] Antoine Bouët and Jeanne Metivier, 'Is the WTO Dispute Settlement Procedure Fair to Developing Countries?' (International Food Policy Research Institute) 6 <www.ifpri.org/publication/wto-dispute-settlement-procedure-fair-developing-countries> accessed 29 April 2023; Malebakeng Forere, 'Revisiting African States Participation in the WTO Dispute Settlement through Intra-Africa RTA Dispute Settlement' (2013) 6 LDR 155, 171–72; Edwin Kessie and Kofi Addo, 'African Countries and the WTO Negotiations on the Dispute Settlement Understanding' <www.trapca.org/wp-content/uploads/2019/09/TWP0807_African_Countries_And_The_WTO_Negotiations_On_The_Dispute_Settlement_Understanding-1.pdf> accessed 15 February 2019.

[77] 'Proposal by the African Group (25 September 2002)' Doc TN/DS/W/1, 5; WTO, 'Special Session of the Dispute Settlement Body—Report by the Chairman (30 January 2015)' Doc TN/DS/26.

[78] Forere, 'Revisiting African States' (n 76) 168–70.

[79] Moon, 'Meaningful Technology Transfer to LDCs' (n 25) 8; Peter Gottschalk, 'Technology Transfer and Benefit Sharing under the Biodiversity Convention' in Hans Henrik Lidgard and others (eds), *Sustainable Technology Transfer: A Guide to Global Aid & Trade Development* (Kluwer Law International 2012) 197; 'Proposal on the Implementation of Article 66.2 of the Trade-Related Aspects of Intellectual Property Rights (TRIPS) Agreement' (n 16).

[80] Article 66.2 uses expressions such as 'shall' that denote strong mandatory obligations and 'should' that denote strong exhortatory commitments.

the 2001 WTO Doha Decision on Implementation-Related Issues and Concerns, which expressly confirms that the provisions of Article 66.2 of the TRIPS Agreement are mandatory.[81] The 2001 Declaration on the TRIPS Agreement and Public Health further reiterated this commitment in paragraph 7.[82] Furthermore, in the course of the procedures, the Dispute Settlement Body panel or the appellate body can easily clarify the relevant concepts because it is part of their mandate to provide authoritative interpretation of the Agreements as stated by Article 3, paragraph 9 of the Rules.[83]

Additionally, the cost of dispute settlement process can be mitigated by making use of Advisory Centre on WTO Law (ACWL).[84] Therefore, there is no justification for African states, the majority of which are LDCs, not to make use of the service that have been already by other countries.[85]

Additional challenges confronting LDCs that ostensibly prevent them from using the DSM are the lack of legal expertise in WTO law and the capacity to collect information concerning trade barriers and opportunities to challenge these, alongside fear of political and economic pressure from developed states.[86] To overcome the lack of legal expertise and capacity to gather and systematize information, experience is drawn from the EU and United States that rely on the private sector and trade associations to provide support to build strong WTO legal cases.[87] Although the private sector in LDCs, is still at its infancy, its involvement in building cases and gathering information in their areas of interest and expertise would have a discernible impact on the ability to lodge successful cases with the DSM. It was also suggested that international development institutions can render additional support in that regard. The pivotal role of the ACWL was once again reiterated, including the existing WTO regional centres such as the Trade Law Centre for Southern Africa (TRALAC).[88]

Some commentators have raised the fear of pressure or retaliation from developed states if a developing state initiates DSM procedures. However, the veracity of this claim has not been established because the meticulous study conducted by Bouët and Métivier did not find any evidence demonstrating that the capacity to

---

[81] WTO, 'Ministerial Conferences—Doha 4th Ministerial Conference—Implementation-Related Issues and Concerns' (n 22).
[82] 'Declaration on the TRIPS Agreement and Public Health (adopted 14 November 2001)' (n 32).
[83] Understanding on the Rules and Procedures Governing the Settlement of Disputes (opened for signature 15 April 1994, entered into force 1 January 1995) 1869 UNTS 401; Taubman (n 75) 142.
[84] Kaushik (n 72) 27.
[85] Forere, 'Revisiting African States' (n 76) 171; Amin Alavi, 'African Countries and the WTO Dispute Settlement Mechanism' (2007) 25 DPR 25, 40.
[86] Victor Mosoti, 'Does Africa Need the WTO Dispute Settlement System?' in Gregory Shaffer, Victor Mosoti, and Asif H Qureshi (eds), *Towards a Development-Supportive Dispute Settlement System in the WTO* (ICTSD 2003) 26 <www.eldis.org/document/A12966> accessed 28 April 2023.
[87] ibid 29–33.
[88] Tralac Trade Law Centre, 'Tralac—Trade Law Centre' (*tralac*) <www.tralac.org/> accessed 9 May 2023.

retaliate by developed states against developing states has an impact on dispute initiation. The threat of economic retaliation by the respondent through either financial aid, preferences granted, or trade does not seem to play a role in the initiation of disputes. In their conclusions, Bouët and Métivier also indicated that states do not even consider their own retaliatory capacity to initiate a dispute.[89]

In conclusion, African LDCs are encouraged to exploit the potential of the DSM to claim the benefits that should accrue to them under Article 66.2. These benefits are jeopardized by the lack of implementation of the obligation to establish the necessary incentives in their territories that could encourage companies and other institutions to transfer technology to LDCs. Using the DSM is not an easy task but with the right determination and focus and by gathering the necessary support in terms of skills, information, and funding, LDCs could witness a breakthrough and eventually see developed states engaging in consultations that could lead to substantive results.

E. Conclusion

On the adoption of the TRIPS Agreement in 1994 there was a strong belief that the establishment of a robust IP system would create an enabling environment for companies located in the developed world to transfer technologies to the developing world, spurring more innovation and thus fostering economic growth in those states. Consequently, Article 67 requires that developed states render technical assistance to developing states and LDCs. The deliberate decision was taken to support the latter states in establishing strong IP legal and institutional frameworks. These initiatives succeeded in establishing stronger IP systems in almost all African states. However, the much-desired technology inflows did not occur and innovation is still wanting.

Nevertheless, the establishment of stronger IP systems was balanced by the obligation imposed by Article 66.2 of the TRIPS Agreement that requires developed states to provide a set of incentives to promote and encourage technology transfer to LDCs by companies and institutions located in their own jurisdictions. This provision has not received much attention, hence its implementation has largely been ignored, thereby denying LDCs the ability to establish the much-desired sound and viable technological base for their development. In the circumstances, developing mechanisms to maximize the implementation of Article 66.2 is one of the most daunting tasks for the TRIPS Council. Therefore, the main focus of the current debates regarding the implementation of this provision concerns the effectiveness of the incentives adopted by developed states to assist LDCs to build the

---

[89] Bouët and Metivier (n 76) 28–48.

sound and viable technological base for their development. In this chapter, proposals are tabled to take full advantage of Article 66.2 of TRIPS so that the provision becomes far more meaningful to benefit LDCs. This includes the proposal for the adoption of guidelines on the monitoring process of the implementation of Article 66.2 in LDCs, the proposal for the establishment of the ACTTI, and the full exploitation of the DSM, including by encouraging the institution of cases at the Dispute Settlement Body related to the lack of compliance with this provision by developed states, with a view to forcing them to establish the necessary incentives in their territories to promote technology transfer to LDC.

PART III
LEVELLING THE PLAYING FIELD TO PROMOTE TECHNOLOGY TRANSFER AND INNOVATION IN AFRICA

6
The Proposal to Establish the Agreement on Trade-Related Issues of Technology Transfer and Innovation (TRITTI)

A. Introduction

In the early 1970s, the United Nations Conference on Trade and Development (UNCTAD) developed a draft International Code of Conduct for the Transfer of Technology to promote liberalization and regulation of trade in technology through the introduction of guidelines on the terms and conditions of transfer of technology for the benefit of developing states. The negotiations that were held from 1976 to 1985 did not yield the intended results due to the divergent views of the negotiating countries. The areas in which stumbling blocks arose are well identified: the most appropriate organization to host and manage the legal instrument, the legal character of the instrument (whether binding or merely in the form of guidelines), the scope and extent of technical assistance to developing states and Least Developed Countries (LDCs), the obligations of developed states, the mechanism to force or persuade the private sector to collaborate in the process of transferring technologies, and the Dispute Settlement Mechanism (DSM).

These challenges turned out to be unsurmountable in the attempt to develop the much-needed unified regulation of the international flows of technologies. Instead, several sectoral regimes were established to regulate flows of technology to developing countries, resulting in an untenable scenario of fragmentation of the international legal framework for technology transfer. Indeed, currently there are more than eighty regional and international agreements that deal disjointedly with technology transfer, with a special focus on environmentally sound technologies.[1] It is however worth noting that all these sectoral regimes have failed to promote satisfactory flows of technologies to developing states, hence the quest for a

---

[1] This includes the United Nations Convention on the Law of the Sea (opened for signature 10 December 1982, entered into force 16 November 1994) 1833 UNTS 397 (hereafter UNCLOS); The Montreal Protocol on Substances that Deplete the Ozone Layer (opened for signature 16 September 1987, entered into force 1 January 1989) 1562 UNTS 408 (hereafter MPSDOL); the Convention on Biological Diversity (opened for signature 5 June 1992, entered into force 29 December 1993) 1760 UNTS 79 (hereafter CBD); and the United Nations Framework Convention on Climate Change (opened for signature 9 May 1992, entered into force 21 March 1994) 1771 UNTS 107 (hereafter UNFCCC) among others.

Innovation in Africa. Fernando dos Santos, Oxford University Press. © Fernando dos Santos 2024.
DOI: 10.1093/oso/9780192857309.003.0006

comprehensive and long-lasting international legal regime, especially from developing countries, still continues.

For some, this regime can be achieved by transforming Locke's duty of charity into a binding legal obligation for it to be effective in assisting developing states to substantiate their claims.[2] Accordingly, the rights-based approach may provide a solid legal foundation for African countries to claim integration into the global innovation system.[3] For the rights-based approach, mere charity is no longer enough; rather, it is crucial to shift the perspective from viewing development as a need, to the stage where it is considered a fully-fledged right.[4] A clear international legal framework is therefore necessary to convert this philosophical idea into a useful tool to encourage technology transfer and innovation for the benefit of developing governments. The propositions illustrated did not provide a concrete proposal on the envisaged legal instrument.

In light of that vacuum and the prevailing challenges that developing states and LDCs face in respect of asserting their right to build a viable and sound basis for their development, a proposal is tabled in this chapter for the adoption of an Agreement on Trade-Related Issues of Technology Transfer and Innovation (TRITTI) within the World Trade Organization (WTO) framework in order to provide certainty on the issue.

It was not within the scope of this book to table a full proposal of the text of the future TRITTI because such a text can only result from global diplomatic negotiations. However, what this book has endeavoured to do is to highlight some of the main topics that must be addressed by the proposed international legal instrument if it is to make any difference in pursuit of effectively promoting the transfer of technology to developing states. The aim is to avoid the same obstacles that brought the previous proposals to a standstill, threatening the success of the current submission. These ought therefore to be tackled first.

The chapter is correspondingly structured to address that objective. Hence, it first engages the issue of scope, which was limited in the UNCTAD Draft Code of Conduct, by imposing obligations solely on developed states and by offering a one-off solution to transfer technology without assessing its subsequent impact. Those issues are reassessed and the scope is expanded as a result. Subsequently, the issue of applicable law is reviewed to propose a cautious approach that can assist

---

[2] Paul Gready, 'Rights-Based Approaches to Development: What Is the Value-Added?' (2008) 18 DevPrac 735, 737; Jaakko Kuosmanen, 'Repackaging Human Rights: On the Justification and the Function of the Right to Development' (2015) 11 JGE 303, 303–20.

[3] Bård-Anders Andreassen, 'On the Normative Core of the Right to Development' (1997) 24 ForumDevStud 179, 182; Peter Uvin, 'From the Right to Development to the Rights-Based Approach: How "Human Rights" Entered Development' (2007) 17 DevPra 597, 598; Andrea Cornwall and Celestine Nyamu-Musembi, 'Putting the "Rights-based Approach" to Development into Perspective' (2004) 25 AmUnivLawRev 1415, 1415–37; Gready (n 2) 737; Kuosmanen (n 2) 303–20.

[4] Aili Mari Tripp, 'Development and the New Rights-Based Approaches in Africa' (2009) 36 ROAPE 279, 279; Brigitte I Hamm, 'A Human Rights Approach to Development' (2001) 23 HRQ 1005, 1026.

in striking the right balance between those who intend that the applicable law be that of the importing countries and those that find it proper to apply the law of the technology-exporting country. The contentious issue of the character of the legal instrument is further examined with the view to determine whether the instrument needs to be mandatory in effect or whether simple guidelines or standards would be an adequate solution. The issue of enforcement and dispute settlement is resolved by taking advantage of the existing DSM established under the WTO, according to the proposal in this chapter. Technical assistance is a common feature in international legal instruments involving developing states and LDCs, hence it too is addressed. The crucial topic of the hosting organization is considered and the WTO is proposed due to its strategic position: it focuses on global trade issues and technology transfer transactions fall within the ambit of trade. Finally, the lack of collaboration from the private sector that owns the technologies to be transferred is one of the main stumbling blocks of any attempt to establish an international legal regime on transfer of technology, hence mechanisms to force or persuade the private sector to collaborate in the process of transfer of technologies are proposed.

In a nutshell, although previous attempts to establish an international regime of technology transfer have failed, the circumstances have changed and the configuration of world geopolitics today seems to be more inclined to gather political support than ever. The main submission of this chapter is that the timing for the TRITTI seems to be ripe and it is worth pursuing without further delay.

B. Rationale for the Establishment of the Treaty

There is a rationale for advocating for the establishment of a new global legal regime based on a multilateral agreement to promote transfer of technology to developing countries.

Philosophically, such a rationale derives from the need to strengthen Locke's duty of charity by establishing a mechanism that can legally compel developed countries to transfer technology to developing countries. To that end, the rights-based approach suggests the empowerment of developing countries and providing full legitimacy to review the current international arrangements through legally based instruments.[5] As suggested by Cornwall and Nyamu-Musembi, the best way of achieving this is to rely in an internationally agreed legal document.[6] These assumptions provide a strong philosophical justification for the adoption of a new international legal instrument to redress the current imbalance in respect of technology issues and to render the duty of charity more effective.

---

[5] Uvin (n 3) 598.
[6] Cornwall and Nyamu-Musembi (n 3) 1415–37.

Secondly, developed states, their innovators, and rights-holders, especially multinational companies, benefited tremendously from the strengthening of intellectual property (IP) rights through the Agreement on Trade-Related Aspects of Intellectual Property Rights (TRIPS Agreement).[7] As a trade-off, LDCs were supposed to benefit from mechanisms such as the TRIPS flexibilities and the commitment to provide trade compensation, including facilitation of technology transfer as provided for by Articles 7 and 66.2. Indeed, facilitating technology transfer for the benefit of developing states forms part of the bargain that persuaded those states to agree to set up strong IP systems.[8] However, it is indisputable that those mechanisms are yet to yield the much-desired results, and LDCs are lagging behind in terms of technology. It is therefore fair to conclude that the current mechanisms to create the balance between protection of IP rights and promotion of technological progress through transfer of technologies are not yet adequate. In addition, the lack of an omnibus legal instrument to address the issue of international technology transfer has prompted the development of sectoral regimes, fuelling the phenomenon of fragmentation. The LDCs therefore need to pursue a more substantive mechanism to achieve that balance and address the current fragmentation in the legal framework on technology transfer.

The difficulties of negotiating a new treaty in the WTO context are acknowledged, especially if the stalemate of the Doha Process in recent years is considered. However, the desire of developed states to maintain the liberalization of trade under the WTO system will certainly leave a window of opportunity for compromises on this issue, something that is long overdue. This is because the most powerful economic blocs such as the European Union and United States (US) cherish the multilateral trading system and view it as a long-term public good, and they would never opt for a system that unilaterally alters the international strategy and that is not premised on respect for multilateral constraints.[9] These two powerful blocks therefore have a vested interest in having the system work properly. Developing states should take advantage of this vulnerability by tabling a well-articulated and serious demand for more equity and balance on technology transfer matters because the likelihood of a compromise with developed states is realistic at the present time. For this reason, in Section C 'Addressing Challenges Raised by the Failed Attempts at Adopting an International Framework', the content of the proposal for the adoption of a multilateral agreement is elucidated. The proposed

---

[7] Mark V Shugurov, 'TRIPS Agreement, International Technology Transfer and Least Developed Countries' (2015) 2 JARE 72, 72–82.

[8] Clemente Forero-Pineda, 'The Impact of Stronger Intellectual Property Rights on Science and Technology in Developing Countries' (2006) 35 ResearchPol 808, 813–14.

[9] Antoine Bouët and Jeanne Metivier, 'Is the WTO Dispute Settlement Procedure Fair to Developing Countries?' (International Food Policy Research Institute) 6–7 <www.ifpri.org/publication/wto-dispute-settlement-procedure-fair-developing-countries> accessed 29 April 2023.

multilateral treaty seeks to persuade developed states to establish suitable mechanisms to promote the transfer of technologies to states in need.

C. Addressing Challenges Raised by the Failed Attempts at Adopting an International Framework

1. UNCTAD Draft Code Stumbling Blocks

If one considers that the UNCTAD Draft Code of Conduct on Technology Transfer failed to gather consensus for its adoption, notwithstanding more than ten years of negotiations, the obvious question to ask is: What will be different with the new proposed TRITTI and what chances of success militate in favour of its adoption?

First and foremost, decades have elapsed since negotiations on the UNCTAD Draft Code stalled in 1985, so, arguably, the circumstances have changed. Given this reality, the proposed TRITTI will be facing a new context that is different from the one prevailing at the time of the failed negotiations. Indeed, the configuration of world geopolitics has drastically changed and the negotiating parties of the Draft Code, being the developed states (Group B), developing states (Group of 77), and the former USSR and its satellite states (Group D) no longer exist or do not exist in their original form. Specifically, Group D is no longer in existence and Group B and the Group of 77, although still extant, have been drastically reconfigured.

Secondly, reviving the debate on an international regime on technology transfer requires that the unresolved issues in the UNCTAD Draft Code of Conduct on Technology Transfer, such as the difference in approach between developed and developing states regarding the binding or voluntary character of the instrument, the definition and scope, the non-inclusion of intra-firm transactions that make up the bulk of technology transfer, the applicable law, and the settlement of disputes, must be tackled before the process can move forward.[10]

Those issues which halted the success of the UNCTAD Draft Code may conceivably be viewed differently now at the negotiating table of TRITTI. The challenges encountered during the negotiation of the UNCTAD Draft Code will be further analysed here, taking into account these new circumstances. Therefore, if the parties adopt a new approach regarding the same issues it may lead to a greater probability of success once negotiations are initiated.

---

[10] Peter Gottschalk, 'Technology Transfer and Benefit Sharing under the Biodiversity Convention' in Hans Henrik Lidgard and others (eds), *Sustainable Technology Transfer: A Guide to Global Aid & Trade Development* (Kluwer Law International 2012) 200; Peter Yu, 'A Tale of Two Development Agendas' (2009) 35 ONULawRev 465, 498; David M Haug, 'The International Transfer of Technology—Lessons That East Europe Can Learn from the Failed Third World Experience' (1992) 5 HarvJL&Tech 221.

2. Scope

The UNCTAD Draft Code proposed the liberalization of trade in technology and the introduction of guidelines on the terms and conditions of transfer of technology to developing states.[11] It was therefore an open-ended treaty that tackled the commercial exchange of all forms of technology. Indeed, the instrument expressly states in paragraphs 1.5 and 2.2 that the Code of Conduct is universally applicable in scope. Regarding the concept of technology, paragraph 1.3 includes a wide spectrum of sectors, such as industrial property, know-how, technical expertise and knowledge for the operation of plants, equipment, intermediate goods, and raw materials.

The TRITTI, proposed in this book, followed the same approach: it is an omnibus treaty when it comes to technology transfer because it is wide in scope in order to embrace all transactions related to transfer of technology occurring worldwide without any limitations. Although the UNCTAD text covered the transfer of all technologies, it was still limited in its scope because it was one-sided, focusing exclusively on the obligation imposed on developed states to transfer technology.[12] The text was also a one-off solution to technology transfer by simply facilitating the transfer of technologies without assessing its subsequent impact.[13] It was submitted that the technology transferred would subsequently stimulate local dynamics, leading to technological progress and enhancing local productive capacity. However, as demonstrated throughout this book, simply facilitating access to technologies is not enough to assist developing states to catch up technologically. Hence, there is a call to move away from the old rhetoric of claiming that more technology should be transferred to Africa, and instead, advocate for the creation of capacities for African states to absorb and assimilate the transferred technology.[14] It is accordingly argued that more than merely providing access to technology, the most important objective of the new international legal instrument should be emphasized, which is to assist states to facilitate technological learning and to develop absorptive capacities and adaptation of technology to the local conditions so that problems relating to local needs can thereby be solved.

Buttressing the argument that a more inclusive process of technology transfer is needed is the knowledge that, in the Draft Code of Conduct, transfer of technology was identified as a one-off mechanism aiming merely at affording access to the technology by developing states and, for this reason, the majority of the efforts

---

[11] UNCTAD, *Transfer of Technology* (United Nations 2001).

[12] Padmashree Sampath and Pedro Roffe, 'Unpacking the International Technology Transfer Debate: Fifty Years and Beyond' (2012) 47–48 <https://papers.ssrn.com/abstract=2268529> accessed 20 June 2021.

[13] Pedro Roffe and Taffere Tesfachew, 'Revisiting the Technology Transfer Debate: Lessons for the New WTO Working Group' (2002) 6 Bridges, ICTSD 7.

[14] Damilola S Olawuyi, 'From Technology Transfer to Technology Absorption: Addressing Climate Technology Gaps in Africa' (2018) 36 JENRL 61, 62.

made in crafting the international legal framework focused on dealing with the glitches identified in the process of transfer of technology at the international level and on the role played by multinational companies.[15] However, transfer of technology cannot be limited to a process of appropriation of individually owned technologies but implies open-ended support to build capacities in developing states. This approach must be clearly addressed in the proposed new treaty. Therefore, the ultimate aim of creating absorptive capacities and technological learning in developing states must feature prominently in the preamble of the TRITTI and in the provisions that will set out the objectives of the legal instrument in order to enable its effective implementation.

A good starting point in that direction is the current undertaking of the Technology Mechanism under the United Nations Framework Convention on Climate Change (UNFCCC) that goes beyond mere technology transfer to encompass support for the beneficiary states to assimilate and absorb technology knowledge, adapt it to local conditions, improve upon it, and disseminate it.[16] Although this is a commendable achievement it is unfortunately still sectoral and limited to the Technology Mechanism.[17] Indeed, the lack of an internationally agreed-upon regime that promotes technology transfer with the ultimate goal of boosting technological capabilities still leaves room for manoeuvre in these sectoral solutions.[18] It is for this reason that the present chapter has advocated the adoption of an all-encompassing international legal regime that would primarily aim at improving the technological capabilities of developing states and LDCs so that technology can genuinely achieve the desired results and be sustainable.

With regard to the most recent international treaties or proposals of an internationally agreed treaty on technology transfer, there are some shortfalls because, generally, they underestimate the overall trade dimension that is embedded in technology transfer. In the case of the proposal tabled by Ma (discussed in Chapter 4), the strong links with the global environmental system led it to focus only on the establishment of a legal instrument that would offer solutions for the transfer of environmentally sound technologies. This limitation is difficult to sustain because the new instrument could capitalize on the rich and substantial content developed and agreed upon in several existing instruments such as the UNCTAD Draft Code of Conduct on Technology Transfer; the Multilateral Agreement on Access to Basic Science and Technology (ABST); the international global environmental protection frameworks, including the CBD, UNFCCC, and its corresponding Kyoto Protocol; and the TRIPS Agreement, to advance a far-reaching instrument. Such an approach would therefore address the fragmentation of the technology

---

[15] Roffe and Tesfachew (n 13) 7.
[16] Sampath and Roffe (n 12) 50.
[17] 'UNFCCC Technology Mechanism' <https://unfccc.int/ttclear/support/technology-mechanism.html> accessed 13 June 2019.
[18] Sampath and Roffe (n 12) 50.

transfer regime that is witnessed today by bringing 'under one roof' all issues related to technology transfer.

Furthermore, it is also worth highlighting that new areas of public interest that were not considered in the negotiations of the UNCTAD Draft Code have emerged—such as health, agriculture (protection of new plant varieties), and climate change—and corresponding autonomous international legal instruments have been devised to address technology transfer in those emerging areas.[19] TRITTI presents a unique opportunity to amalgamate within one instrument all of the solutions that were incorporated in those treaties. This would be an easy exercise because the provisions of those legal instruments were fully negotiated and agreed to by the parties. The provisions of TRITTI must therefore be crafted not only with a view to covering the new areas highlighted, but also to cater for any emerging areas in the future.

3. Applicable Law

The positions of developed states and developing states on the law applicable to technology transfer transactions also differ substantially. Developing states that usually import technology demand the application of their own law. Arguably, developed states that are the main exporters of technology prefer the application of general principles of private international law, and in particular the principle of party autonomy, that leads to the application of the law of the technology-exporting state.[20] This is because international law on technology transfer can only complement national law. Indeed, since transactions related to technology transfer largely take place between private firms, the regulation of these transactions is effected through domestic contracts, competition, and property law which, by their very nature, are territory-dependent.[21]

The fundamental question to be answered in this context is therefore which national law applies to technology transfer? Is it the national law of the exporting or of the importing state? A consideration of the history of this question is necessary. At the close of the sixth session of the UN Conference on the Draft Code of Conduct on Transfer of Technology on 5 June 1985 no consensus had been reached on the applicable law. The Group of 77 favoured the application of the law of the recipient state on the basis that it distinguished between matters of public policy (*ordre public*) to which the laws of the recipient state would apply, and matters pertaining to private interests, where the choice of applicable law would be left to the

---

[19] ibid.
[20] Michael Waibel and William P Alford, 'Technology Transfer' in R Wolfrum (ed), *Max Planck Encyclopedia of Public International Law* (Oxford University Press 2011) 801–14.
[21] ibid.

parties.[22] In contrast, Groups B and D favoured the application of rules of private international law. A compromise text suggested that parties would define the applicable law by common consent, except when it came to national rules that cannot be derogated from by contract.[23]

In regulating this issue, the proposed TRITTI must be cautious in striking the right balance. Indeed, any attempt to force developed states to apply the laws of importing states in order to regulate transfer of technology will be viewed with scepticism, and suspicion may derail efforts to reach an agreement on an international legally binding instrument. On the other hand, developed states may only be assured that technology transfer is taking place on terms that can benefit them if their national law is applied. For the purposes of this book, the solution lies in between the two opposite poles of the spectrum and consists in crafting laws in both exporting and importing states that may create an enabling environment to promote technology transfer. This is achievable through enacting national laws in developed states that provide the right incentives to cause local companies to engage in technology transfer for the benefit of developing states under favourable conditions. On the part of developing states, laws must provide the assurance that protection of the technology exported to the state is guaranteed by way of robust intellectual property rights, safeguarding contracts on technology transfer and protection of investments. There is established practice that follows this pattern in the most recent international sectoral instruments that deal with international technology transfer. For example, the Convention on the Law of the Sea encourages member states to actively cooperate on marine technology transfer on 'fair and reasonable terms and conditions' but safeguards existing proprietary rights and exhorts states to 'foster favourable economic and legal conditions for the transfer of marine technology for the benefit of all parties concerned on an equitable basis', balancing the rights and duties of holders, suppliers, and recipients of marine technology.[24]

To prompt the exporting and importing nations to enact such national provisions, TRITTI should set them as an obligation that must be complied with by the members that will accede to the Treaty. The existence of such a set of laws on both sides that take into account the specificities of each side and seek to achieve different objectives is the correct approach to address the concerns of the parties in ascertaining which of the applicable laws applies.

---

[22] Michael Blakeney, *Legal Aspects of the Transfer of Technology to Developing Countries* (ESC Pub 1989) 158–59.
[23] ibid 159.
[24] Arts 266 and 267 of UNCLOS (n 1).

4. Legal Character of TRITTI

The main reasons behind the failure of the UNCTAD Draft Code were the divergent positions regarding the legal character of the instrument. Developing states consistently pushed for mandatory effect of the instrument as evidenced by the text of paragraph 13 of the preamble, which expressly states that the parties agree on the adoption of an internationally legally binding code of conduct on the transfer of technology. Conversely, developed states pursued guidelines or standards to be observed by the companies involved in technology transfer.[25] Those intransigent positions on this issue were at the heart of the failure of the UNCTAD Draft Code.

The Convention on the Law of the Sea initially followed the same trend and through the combined provisions contained in Article 144 and Annex III entitled 'Transfer of Technology' the transfer of marine technology was made mandatory, with specific regard to the seabed mining-associated technology.[26] With such provisions in place, western states, led by the US, long resisted acceding to the Convention. The deadlock threatening the survival of the Convention was only overcome in 1994 through UN Resolution 48/263 ('Agreement relating to the Implementation of Part XI of the United Nations Convention on the Law of the Sea of 10 December 1982'),[27] which repealed Article 5(3) that had imposed the mandatory regime.

The lesson to be learned from this development is that it is unrealistic to impose legally binding instruments and expect that developed states will sign up to them. Taking this type of stance inevitably condemns the initiative to premature failure. Therefore, the proposed TRITTI has a greater chance of success if it follows the more malleable approach introduced by the amended Convention on the Law of the Sea. Notable also is the fact that although other recent legal instruments have pursued the 'binding instrument' route, as in the case of the UNFCCC, the Kyoto Protocol, and other international treaties that address issues related to transfer of technology, they have achieved limited results in promoting the transfer of technology. Apparently, the main focus should not be on the legal character of the instrument but rather on how to make the provisions enforceable among states and persuasive to the private sector. The use of the WTO DSM will provide part of the solution. In addition, linking technology transfer to trade and regulating it through the specific type of national laws as subsequently analysed also provides a possible answer to this concern. Paradoxically, it has been observed that there are other stipulations on technology transfer that are not legally binding but provide

---

[25] In the proposed text of the preamble by the developing countries it was clearly stated that 'an internationally legally binding instrument is the only form capable of effectively regulating the transfer of technology' hence such instrument was strongly recommended for adoption, see Blakeney (n 22) 134–35.
[26] James Stavridis, 'Marine Technology Transfer and the Law of the Sea' (1983) 36 NWCR 38, 41.
[27] UNCLOS (n 1).

better solutions and are more persuasive, hence more apt to promote technology transfer.[28] The Bali Action Plan, the Copenhagen Accord, and Agenda 21 (United Nations Sustainable Development Goals), are some of the examples of instruments that are not legally binding but are viewed as more persuasive in the negotiations regarding technology transfer.[29]

Notwithstanding the cogency of the argument that a formal, legally binding instrument is the best solution, practice seems to suggest that such a radical position is untenable and counterproductive. In this regard, negotiations should not focus on the legal character of TRITTI, but rather on establishing the most effective and appropriate mechanism to enable its enforcement and to be able to persuade the private sector to collaborate in the transfer of technology. As such, Section C.5 on 'Dispute Settlement' addresses the most viable enforcement mechanism; subsequent sections deal with the collaboration of the private sector.

5. Dispute Settlement

In the context of the debates on the Draft Code on Technology Transfer two opposing positions were tabled: one from developing states suggesting that the state acquiring the technology should have exclusive jurisdiction over any disputes arising therefrom; and the other from developed states advocating that the parties should be free to choose the preferred forum for the purpose of resolving disputes.[30] The two groups later converged on the idea that disputes should be amicably resolved or that there should be recourse to arbitration to settle them. That position prevailed and has remained the essence of the content of Article 9 of the Draft Code.[31] In this regard, in the context of the ABST, the initial proposal to adopt an ABST envisaged that the arrangement should be placed within the broad jurisdiction of the WTO. This format would have made it easier, at least for the disputes arising between governments related to the transfer of technology, for such disputes to be settled under the existing WTO DSM.[32] Nonetheless, the ABST mechanism was never operationalized and therefore that format did not materialize.

The most recent sectoral Multilateral Environmental Agreements that address issues related to technology transfer also offer a variety of solutions for the

---

[28] Zhong Fa Ma, 'The Effectiveness of Kyoto Protocol and the Legal Institution for International Technology Transfer' (2012) 37 JTT 75, 91.
[29] 'Bali Action Plan (adopted 14 March 2008)' Doc FCCC/CP/2007/6/Add.1; 'Copenhagen Accord (adopted 30 March 2010)' Doc FCCC/CP/2009/11/Add.1; 'Agenda 21: Sustainable Development Knowledge Platform' <https://sustainabledevelopment.un.org/outcomedocuments/agenda21> accessed 19 June 2019.
[30] Blakeney (n 22) 159.
[31] ibid 160.
[32] John Barton and Keith Maskus, 'Economic Perspectives on a Multilateral Agreement on Open Access to Basic Science and Technology' (2004) 1 SCRIPT-ed 379.

settlement of disputes. Article 14 of the UNFCCC establishes that the settlement of disputes arising from the interpretation of the Convention shall be effected through negotiation or any other peaceful means of the parties own choice. The disputes may also be submitted to the International Court of Justice and or to arbitration, according to Article 14.2(a) and (b). Article 19 of the Kyoto Protocol to the UNFCCC endorses these provisions and makes them applicable to the Protocol. Similarly, the 1992 Convention on Biological Diversity also deals with the settlement of disputes in Article 27, and what was chosen for the settlement of disputes related to the Treaty was negotiation, mediation, arbitration, conciliation, and submission of cases to the International Court of Justice. Therefore, the current scenario with regard to the settlement of disputes related to technology transfer is a fragmented one. Accordingly, TRITTI could capitalize on the current chaos by submitting all possible conflicts arising from technology transfer transactions to the existing WTO DSM. This is recommended because the WTO system establishes a mandatory and unified tool that can be used in any trade disputes under all WTO agreements.[33] Appendix 1 of the Understanding on Rules and Procedures Governing the Settlement of Disputes (the Understanding) includes among the agreements within its remit the TRIPS Agreement. According to the Understanding, member states can lodge complaints to the WTO DSM related to three kinds of disputes, namely violation complaints, non-violation complaints, and situation complaints.[34] Therefore, based on Article 23 of the Understanding, a member state can lodge a complaint if any benefit accruing to it either directly or indirectly is being nullified or impaired, or the attainment of any objective of the Agreement is being impeded as a result of the failure of another contracting party to carry out its obligations. Thus, any violation of a mandatory obligation by any WTO member may trigger a complaint by another member.[35]

Consequently, a distinct advantage of establishing TRITTI under the WTO system is the possibility of having recourse to the DSM whenever conflicts arise. Governments could commit to supporting owners of technologies and candidate recipients of those technologies whenever they decide to make use of the DSM provided under the WTO. Notably, the main reason behind the proposal to establish TRITTI under the WTO is precisely to take advantage of the DSM. Indeed, although UN-based frameworks, such as the CBD, UNFCCC, and Kyoto Protocol contain binding provisions, they have completely failed to promote technology transfer due to the lack of an enforcement mechanism.[36] This challenge is

---

[33] Monika Bütler and Heinz Hauser, 'The WTO Dispute Settlement System: A First Assessment from an Economic Perspective' (2000) 16 J LawEconOrgan 503, 509; Malebakeng Agnes Forere, *The Relationship of WTO Law and Regional Trade Agreements in Dispute Settlement: From Fragmentation to Coherence* (Wolters Kluwer: Kluwer Law International 2015) 128–29.

[34] Bütler and Hauser (n 33) 509; Forere (n 33) 128–29.

[35] Xavier Groussot and Thu-Lang Tran-Wasescha, 'TRIPS Article 66(2)—Between Hard Law and Soft Law?' in Hans Henrik Lidgard, Jeffrey Atik, and Tu Thanh Nguyen (eds), *Sustainable Technology Transfer: A Guide to Global Aid & Trade Development* (Kluwer Law International 2012) 19.

[36] Ma (n 28) 83–88.

addressed through the establishment of the new regime under the WTO umbrella enabling the application of the DSM.

6. Technical Assistance

In order to capacitate developing states and LDCs to access, use, absorb, and adapt technology to their own needs, adequate support from developed states is required.[37] The UNCTAD Draft Code alludes to supporting these states in Chapter 6 under the heading 'Special treatment to developing countries'. Nevertheless, since the Code was never adopted it did not lead to any concrete implementation. Instead, it is Article 67 of TRIPS that has required developed states to provide the necessary technical and financial cooperation to developing states and LDCs that has prompted several initiatives. Paragraph 2 of article 67 is explicit on the type of support to be rendered to developing states and LDCs. This includes assistance in the preparation of laws and regulations on the protection and enforcement of IP rights as well as on the prevention of abuse of those rights, and the establishment or reinforcement of domestic offices and agencies relevant to these matters, including the training of personnel. The subsequent action undertaken by the developed world to fulfil these obligations has prompted the development of strong legal and institutional IP frameworks in developing states and LDCs. Apparently, developed states have been promoting the uptake of the IP system not as tool for development, but as a means of safeguarding the IP rights of their own companies that do business abroad. The objective of promoting technological progress, innovation, and economic development in foreign states, especially in LDCs, appears to have been held in abeyance and replaced by the strengthening of the IP system as an end in itself. To counter that trend, it is proposed that technical assistance should henceforth be developed as part of the overall economic, industrial, science and technology, innovation, and business sector development strategy.[38] To that end, technical assistance must be geared towards fostering local innovation, taking into account the national context, as well as the needs and expectations of developing states. For example, it was observed from the reports submitted to the TRIPS Council by developed states under Article 66.2 that many projects are serving trade policy needs rather than innovation policy needs and are therefore not likely to improve the economic situation of the poorest states.[39] It follows that the issue of technical assistance needs to be expressly included in the future TRITTI, and

---

[37] Chantal Thomas, 'Transfer of Technology in the Contemporary International Order' (1998) 22 FordhamIntLJ 2110.
[38] Carolyn Deere-Birkbeck and Ron Marchant, 'The Technical Assistance Principles of the WIPO Development Agenda and Their Practical Implementation' (2010) Issue Paper No 28, 13.
[39] Dominique Foray, 'Technology Transfer in the TRIPS Age: The Need for New Types of Partnerships between the Least Developed and Most Advanced Economies' [2009] IPSDS 51–56.

concrete mechanisms must be established for its operationalization. In particular, it is imperative that the preamble to TRITTI and provisions setting out its objectives clearly state the crucial role of technology transfer as an effective tool for innovation and growth in LDCs and developing states in general. This will guide all other initiatives towards achieving this important objective. This is possible if the new terminology and approach that have been adopted of late are developed in the international arena, such as in Article 67 of TRIPS, and if Article 4 of the UNFCCC is taken into account as the foundation of the new regime. For instance, the terminology used by Article 4.5 of the UNFCCC is compelling and expressly calls upon developed states to 'take all practicable steps to promote, facilitate and finance, as appropriate, the transfer of, or access to, environmentally sound technologies and know-how to other Parties'. Most importantly, the Convention urges those states to support the development and enhancement of endogenous capacities and technologies of developing states parties to the Convention. Undoubtedly, the TRITTI provisions to be proposed must therefore capitalize on the terminology in use in the recent sectoral regulations of technology transfer that are focusing in developing absorptive and adaptive capabilities of developing states in order to fast-track technological catch-up and innovation.

7. WTO as a Host Organization

During the negotiations of the Draft Code of Conduct it was agreed that the institutional machinery of the Code of Conduct on Technology Transfer would be provided by UNCTAD (a United Nations Agency). UNCTAD would therefore act as a Secretariat for the institutional body according to Article 8.4 of the Draft Code. This issue was not contentious and both parties were unanimous in entrusting UNCTAD with the administration of the legal instrument, although with slight differences of opinion on the best mechanism of operationalization.[40] Due to the failure of the UNCTAD Draft Code, the institutional machinery proposed never took off. Specifically, the fragmentation of the technology transfer regime that followed also prompted the fragmentation of the institutional framework. Several institutions were entrusted with the administration of sectoral technology transfer regimes. For example, in 2001 the Conference of the Parties (COP) within the context of the UNFCCC and the Kyoto Protocol established an Expert Group on Technology Transfer (EGTT) with the objective of facilitating and advancing technology transfer activities and making recommendations to the Subsidiary Body for Scientific and Technological Advice (SBSTA) and the Subsidiary Body for Implementation (SBI). The EGTT contributed very little in promoting technology

---

[40] Blakeney (n 22) 156–58.

transfer, partly because of the lack of coercive power of the SBSTA.[41] Nor did other international legal instruments, especially the multilateral environmental agreements, yield tangible results with regard to technology transfer. All of this suggests that the UN framework is inadequate to deal with the issues of technology transfer because its approach leans more toward issues of public interest, whereas technologies are owned by the private sector that is primarily guided by the objective of profit-making. Since the UN frameworks do not involve the private sector in their processes and do not pursue those interests, it is unlikely that a system developed in that context will adequately address them.[42]

The WTO is therefore unique in that it focuses on global trade issues, and it has developed the TRIPS Agreement, establishing the minimum standards for the protection of IP rights, including Articles 7, 8, and 66.2 that deal with technology transfer, and has devised a unique DSM. This institution seems to be the best forum to administer the proposed international legal instrument because it contains adequate features to enable implementation of the Treaty.

The WTO platform is appropriate because technology transfer transactions are in the nature of trade, and therefore efforts to facilitate transfer of technologies will always affect trade and have an impact on the trade system that is now governed by the WTO. This is the case with respect to incentives, subsidies, or any other initiatives that may give special treatment to developing states and LDCs. Further, the fact that the WTO is also a negotiating forum on trade issues may facilitate trade-offs in concessions across sectors and agreements that may have an impact on technology transfer. TRITTI and its institutional framework could leverage its strategic position at the WTO to effectively participate and influence the debates running in parallel in the WTO TRIPS Council regarding technology transfer. The WTO could be the right location to set the balance between the protection of IP rights and promoting access, use, absorption, and adaptation of technology by developing states. TRITTI could also be an excellent mechanism to accelerate implementation of Article 66.2 of TRIPS and promote technology transfer. By way of illustration, annual reports presented by developed states on the incentives established under Article 66.2 seem to suggest that some flows of technology directed to LDCs are occurring. However, it is questionable whether those flows are really an expression of technology transfer that can promote innovation in LDCs. Since the objective of TRITTI will not be limited to facilitating technology transfer transactions exclusively, but also trying to effectively create an enabling environment for promoting technological catch-up, the TRITTI machinery could assist in the measurement of the impact of the global implementation of Article 66.2 of TRIPS. The TRITTI framework could therefore be a focal point for monitoring the implementation of the commitments undertaken under the TRIPS Agreement on

---

[41] Ma (n 28) 81.
[42] ibid 92.

technology transfer and evaluating effectiveness in promoting technology transfer and the development of technological capabilities of recipient states.

8. Measures to Promote Collaboration of the Private Sector in the Transfer of Technologies

One of main challenges in the establishment of an effective regime to promote technology transfer at the international level is the lack of collaboration from the private sector that owns the technologies to be transferred. Philosophically, Gordon recognized the rights of the creators to their own endeavours provided that the ownership does not deprive others of the benefit of existing knowledge.[43] Nevertheless, Gordon did not provide guidance on how the limitation of those rights ought to be effected and how sharing of their endeavours is to be enforced.

The UNCTAD Draft Code of Technology Transfer also failed because it did not address this issue appropriately as it was anchored in the establishment of obligations to governments of the member states only, which is challenging and not effective since technologies are owned by private entities. Indeed, developed states have always expressed their inability to compel the transfer of technology, claiming that they do not own the vast majority of technologies subject to transfer and cannot force the private sector to transfer technologies.[44] Consequently, those states argued that incentives can only take the form of encouragement, promotion, and facilitation of projects that promote transfer of technologies.[45] This persistent position of developed states seems to be a clear sign of market failure. However, as much as the arguments of developed states have some legitimacy, mechanisms to encourage those entities to transfer technology to developing states and LDCs ought to be sought. Redressing this market failure is possible but requires government intervention. The idea of a 'capable state' that has of late been advocated by many international organizations, including the World Bank, seems appropriate in this context. A 'capable state' has the ability to trigger dynamics that can stimulate initiatives in the private sector, complement private investment whenever necessary, and enhance cooperation and coordination with and within the private sector.[46] Therefore, for the TRITTI to be successful it should establish mechanisms

---

[43] Wendy J Gordon, 'A Property Right in Self-Expression: Equality and Individualism in the Natural Law of Intellectual Property' (1993) 102 YaleLJ 1533, 1540.

[44] European Union, 'Report on the Implementation of Article 66.2 of the TRIPS Agreement' (WTO) Doc IP/C/R/TTI/EU/3 <https://docs.wto.org/dol2fe/Pages/FE_Search/FE_S_S009-DP.aspx?language=E&CatalogueIdList=288475,288461,288378,288403,288245,288236,288039,288037,288041,279272&CurrentCatalogueIdIndex=0&FullTextHash=&HasEnglishRecord=True&HasFrenchRecord=True&HasSpanishRecord=True> accessed 29 October 2023.

[45] ibid.

[46] Joachim Ahrens, 'Governance and the Implementation of Technology Policy in Less Developed Countries' (2002) 11 EINT 441, 442–44.

that are capable of persuading the private sector to collaborate with their respective governments to facilitate the transfer of technologies that they own to developing states. In order to be able to persuade the private sector to adhere to the new regime, the legal instrument must be linked to trade issues and must have an enforcement mechanism that will safeguard private IP rights. It is therefore clear that the new regime must be anchored to a framework underpinned by the IP system and in an organization that can authoritatively pronounce itself on trade matters. This leads us, unequivocally, to the WTO, the only global international organization dealing with the rules of trade between nations that was established outside the UN system to ensure that trade flows as smoothly, predictably, and freely as possible.[47]

At the operational level, it is proposed that TRITTI should contain provisions that address the concern of promoting the 'collaboration of private sector in the process of transfer of technologies'. This could be in the form of general commands or minimum standards that would be imposed on the member states to enact and implement national legislation that is conducive to the transfer of technology. One way of achieving this is through the introduction of requirements imposed upon all states that are members of TRITTI to establish regulatory benefits for the private sector in the context of the national trade regime in exchange for their facilitating the transfer of technologies. For example, if both exporting and importing states are bound by TRITTI to enact national laws that offer incentives to companies that are engaging in technology transfer to developing states under favourable conditions, this would certainly make a difference. Relevant, in the context of the implementation of Article 66.2 of the TRIPS Agreement, is the proposal tabled by the LDCs to the TRIPS Council for developed states to set conditions for companies in their jurisdictions to participate in contracts tendered for by their governments in exchange for incentives to those companies to transfer technology. In these circumstances, LDCs accept that developed states could provide for the payment of royalties as an incentive when technologies are effectively transferred to LDCs by enterprises in developed states.[48]

Developing states should also play a role through the enactment of national laws that provide assurances that technology exported to their states is safe by way of protection of IP rights, safeguarding the contracts of technology transfer and protection of the investments. This would make it possible for private companies to move their technologies confidently, with the assurance of fair returns deriving from respect for their rights in the new markets. There is a good precedent in the

---

[47] See <www.wto.org>.
[48] 'Proposal on the Implementation of Article 66.2 of the Trade-Related Aspects of Intellectual Property Rights (TRIPS) Agreement—Communication from Cambodia on Behalf of the LDC Group (WTO 2018) Doc IP/C/W/640 <https://docs.wto.org/dol2fe/Pages/FE_Search/FE_S_S009-DP.aspx?language=E&CatalogueIdList=243589,243337,243336,243182,243183,243179,243200,241809,240388,239456&CurrentCatalogueIdIndex=6&FullTextHash=&HasEnglishRecord=True&HasFrenchRecord=True&HasSpanishRecord=True> accessed 12 March 2019.

WTO of an arrangement that has persuaded states to establish trade-related effective systems. One such example is the TRIPS Agreement itself, which catapulted the IP system to its highest level by requiring all members of the WTO to establish legal and institutional frameworks for the administration and enforcement of IP. For example, with regard to enforcement, Article 41 of TRIPS requires members to ensure that enforcement procedures are available under their law so as to permit effective action against any act infringing on IP rights covered by the Agreement. This type of requirement may also be included in the proposed TRITTI with regard to the necessary collaboration of the private sector in the international transfer of technologies in exchange for enhancement of favourable conditions of trade in the beneficiary states.

As observed by Ma, although the TRIPS Agreement has attempted to introduce some regulations to promote technology transfer, such as Article 66.2, it has not been effective because it focused on regulating technology transfer from industrialized states to the least developed states, thus excluding transfer of technologies to states that are not deemed 'least developed' and also excluding the responsibility of developing states and LDCs.[49] Based on this, Ma proposed the adoption of an international Technology Transfer Treaty beyond (and in addition to) TRIPS. Nevertheless, the scope of Ma's proposal is very narrow. As a result, a proposal to adopt an omnibus TRITTI under the WTO appears to be an opportunity both to expand the narrow scope proposed by Ma and to address all the inefficiencies that were identified in the international technology transfer system currently operating worldwide.

D. Conclusion

The rights-based approach to development is a powerful tool that can operationalize the right to development by enabling developing states to change the duty of charity that was formulated by Locke into a legal obligation.[50] An important component of development is technological progress and innovation.[51] However, this important element is threatened by the lack of a clear legal regime at the international level to enable all states to benefit. Attempts to overcome this challenge have failed or had limited success, as evidenced by the case of the failure to adopt the UNCTAD Draft Code of Conduct on Technology Transfer in 1985. The absence of such a unified legal regime left room for separate provisions that were

---

[49] Ma (n 28) 93.
[50] Arjun Sengupta, 'On the Theory and Practice of the Right to Development' (2002) 24 HRQ 837, 846; Tripp (n 4) 279; Uvin (n 3) 598.
[51] Engwa Azeh Godwill, 'Science and Technology in Africa: The Key Elements and Measures for Sustainable Development' (2014) 14 GJSFR: G BIO-TECH & GEN 16, 21; Stephen Young and Ping Lan, 'Technology Transfer to China through Foreign Direct Investment' (1997) 31 RegStud 669, 670.

devised by several international legal instruments, especially in the environmental arena, leading to fragmentation and consequently rendering such frameworks ineffective.

Based on the submissions on the issue, the suggestion in this book is to fill the gap in the legal framework by proposing the adoption of a uniform international technology transfer agreement under the framework of the WTO that could be designated as the Agreement on Trade-Related Issues of Technology Transfer and Innovation (TRITTI). The new instrument would bring finality to the debate regarding the need to create the right balance between the protection of IP rights of the developed world and facilitating access to technology with a view to promoting technological progress to the less privileged states. The timing for the TRITTI seems to be ripe as several other legal instruments have set undisturbed sectoral regimes of transfer of technology that can be mainstreamed into this omnibus global instrument. Its value proposition seems to be more appealing: an instrument that will establish a mechanism that will not only create international obligations among states but will be cascaded down to the member states through concrete provisions in national legislation that will encourage the private sector to act. Emphasis will accordingly be placed on creating absorptive capabilities that will enable developing states to access, assimilate, adapt, and use technology for their own benefit. Therefore, an international technology transfer treaty is conceivably likely to gather political support today, more than ever before, and it is worth pursuing such a treaty without further delay.

7
The Role of Developing States and LDCs in the Quest for Technology Transfer: Recommendations

A. Introduction

The premise of this book is the prevailing assumption that, for Africa to reclaim its space in the global innovation arena, external factors must be considered. This view has been deceptive since it gives the impression that encouraging technology transfer is a one-way process that is only designed to compel developed states to enable flows of technologies to LDCs, and it therefore needs to be revised. The Agreement on Trade-related Aspects of Intellectual Property Rights (TRIPS Agreement), being the main international legal instrument that has shaped the intellectual property (IP) system in recent times, and which has a bearing on technology transfer and innovation, is central to the analysis.

Technology transfer also requires proactive efforts by LDCs in trying to set up an enabling environment through the establishment of the right mechanisms and incentives to draw technologies into their respective states. This is the case because encouraging technology transfer may be greatly aided by host-state incentives and a favourable enabling environment. Furthermore, in Chapter 6, one of the recommendations made regarding the contents of the Agreement on Trade-Related Issues of Technology Transfer and Innovation (TRITTI) was the obligation of developing states or LDCs to attract and absorb technologies. This chapter expands on that recommendation, with a view to elaborating on the role that developing states and LDCs can play in attracting and absorbing technology. This is important because the literature and international endeavours concentrate mainly on the role of developed states in transferring technology and less so on the role of recipient states. More accurately, the existing literature points out that some of the interventions that states can undertake to promote flows of technologies include the liberalization of regulatory frameworks; improvement of national policies; amelioration of the competitive atmosphere, including in the cultural and socio-economic environment; protection of IP rights; development of an adequate infrastructure, including industrial structures; and the expansion of scientific institutes, research and development centres, training institutes, and provision of skilled personnel.[1] Other specific incentives proposed to spur the flow of technology include direct subsidies, tax preferences, trade restrictions, differentiation on the inflows of foreign investment, credit subsidization, and financial restraints.[2]

Besides attracting foreign technologies, the capacity of recipient states to absorb the technology transferred is of utmost importance. Indeed, it has been established that the ability to exploit external knowledge is one of the most important components of innovation capabilities.[3] East Asian countries are known to have moved from being technology-averse states to fast-developing states, owing to their ability to imitate, absorb, and assimilate foreign technologies.[4] This chapter therefore determines the role that African states can play in order to attract and absorb the technology that developed states are able to transfer. In so doing, the chapter draws lessons from the successful experiences of East Asian states that have managed to overcome a technology-averse culture and that have enabled innovation and technology transfer to thrive.

The chapter is organized as follows: the first section focuses on the development of technology absorption and adaptation capabilities as enablers of technology transfer in Least Developed Countries (LDCs). Accordingly, the absorptive capacities of states are reviewed and measures to develop them analysed. These include: the establishment of institutional frameworks focusing on science, technology, and innovation; improving funding for research and development (R&D); development of human capital to enable technological development; and the adequate exploitation and socialization of science, technology, engineering, and mathematics. The chapter then delves into the particular role of utility models as a means to facilitate technological learning, improvement of absorptive capacities, and adaptation of imported technologies to local needs. A example of a national strategy to promote the use of utility models is provided. The chapter closes by hinting at the need to fully exploit the opportunities offered by the digital economy and the Fourth Industrial Revolution (4IR) and accordingly it is recommended that governments prioritize human skills development, investment in infrastructure, and funding.

The underlying premise of this chapter is that transfer of technologies from developed to developing countries is not a panacea for technological progress

---

[1] Stephen Young and Ping Lan, 'Technology Transfer to China through Foreign Direct Investment' (1997) 31 RegStud 669, 670.

[2] Joachim Ahrens, 'Governance and the Implementation of Technology Policy in Less Developed Countries' (2002) 11 EINT 441, 451; Amy Jocelyn Glass and Kamal Saggi, 'International Technology Transfer and the Technology Gap' (1998) 55 JDE 369, 369–98.

[3] Wesley M Cohen and Daniel A Levinthal, 'Absorptive Capacity: A New Perspective on Learning and Innovation' (1990) 35 ASQ 128, 128.

[4] Nagesh Kumar, 'Intellectual Property Rights, Technology and Economic Development: Experiences of Asian Countries' (2003) 38 EPW 209, 213; Linsu Kim, 'Technology Policies and Strategies for Developing Countries: Lessons from the Korean Experience' (1998) 10 TechnoAnalStrateg 311, 316–18.

in Africa. Rather, more needs to be done by the African countries themselves to absorb and exploit the technologies transferred, which calls on them to be more proactive.

B. The Development of Technology Absorption and Adaptation Capabilities as Enablers of Technology Transfer in LDCs

1. Development of Absorptive Capacities in Africa

An essential component of the process of technology transfer is the capability of absorbing the technology transferred.[5] The seminal article by Cohen and Levinthal defines absorptive capacity as the capability to recognize the value of new knowledge, to assimilate it, and to apply it to commercial ends.[6] Absorptive capacity consists of four primary capabilities, namely acquisition, assimilation, transformation, and exploitation.[7] Acquisition capability is the firm's ability to identify and acquire beneficial knowledge. This usually takes place through interaction with other firms. Assimilation capability is the ability to analyse and understand new knowledge in the context of the firm and to determine whether it fits into existing knowledge or whether it must be altered to fit into the firm's structures. Transformation capability enables the firm to modify, adapt, and combine new knowledge with existing knowledge, while exploitation capability enables the firm to leverage knowledge or change it in order to create new goods, processes, or organizational forms.[8] Although it has been argued that the four capabilities are complementary, it is the exploitation capability that holds the potential to generate value for the firm which confers competitive advantage.[9] Similarly, to be able to exploit the knowledge requires the capacity to evaluate and utilize the knowledge acquired, which is largely a function of the level of prior related knowledge possessed.[10] That

---

[5] Carsten Gandenberger and others, 'Factors Driving International Technology Transfer: Empirical Insights from a CDM Project Survey' [2015] ClimPol 1068; World Bank, 'Global Economic Prospects 2008: Technology Diffusion in the Developing World (Vol. 2)' (World Bank 2008) 2–14 <http://documents.worldbank.org/curated/en/827331468323971985/Global-economic-prospects-2008-technology-diffusion-in-the-developing-world> accessed 29 October 2023.

[6] Cohen and Levinthal (n 3) 128; Gergana Todorova and Boris Durisin, 'Absorptive Capacity: Valuing a Reconceptualization' (2007) 32 AMR 774, 774–86; Wesley Cohen and Daniel Levinthal, 'Fortune Favors the Prepared Firm' (1994) 40 ManageSci 227, 227–51.

[7] Shaker A Zahra and Gerard George, 'Absorptive Capacity: A Review, Reconceptualization, and Extension' (2002) 27 AMR 185, 189–90; Joshua J Daspit and Derrick E D'Souza, 'Understanding the Multi-Dimensional Nature of Absorptive Capacity' (2013) 25 299, 300–02; Kevin Zheng Zhou and Fang Wu, 'Technological Capability, Strategic Flexibility, and Product Innovation' (2010) 31 SMJ 547, 547–61; Cohen and Levinthal (n 3) 128–52.

[8] Daspit and D'Souza (n 7) 301–02.

[9] Cohen and Levinthal state in this regard that '[a]bsorptive capacity refers not only to the acquisition or assimilation of information by an organization but also to the organization's ability to exploit it', Cohen and Levinthal (n 3) 128; Daspit and D'Souza (n 7) 301–02.

[10] Cohen and Levinthal (n 3) 128.

prior knowledge includes basic skills or scientific or technological knowledge in a given field.

The concept of absorptive capacities transcends the firm level and it is also applicable to states because the state's absorptive capacity is dependent on the absorptive capacities of the organizations in its territory.[11] This can best be illustrated by the example of East Asian states such as Hong Kong, India, Singapore, South Korea, and Taiwan that have benefited immensely from inward technology transfer because their national innovation systems focused on strengthening their national absorptive capacities during the thirty-year period between 1950 and 1970.[12] The absorptive capacities were developed through deliberate investments in scientific and technical training, and as a result of economic policies that enabled companies to compete freely. Public and private institutions that funded and undertook R&D, and further exploited the results for commercial purposes, were all considered part of the National Innovation Systems that were geared towards creating absorptive capacities in these states.[13]

Developing states, especially in Africa, are experiencing difficulties related to the poor quality of their absorptive capacities and, as a result, are struggling to attract and retain foreign technologies.[14] Therefore, government intervention through appropriate policies is required, to enable owners of technologies located in developed states to share their technologies with developing states. The main challenges that African states are facing to develop their absorptive capacities are highlighted and possible interventions to overcome those challenges tabled.

2. Government Interventions to Develop Absorptive Capacity

Acquisition of technology to be infused in barren soil will not ignite innovation. Indeed, there are basic requirements necessary to enable the technology transferred to bear fruit.[15] It is for this reason that scholars have emphasized those elements necessary to facilitate the acquisition of a sound and viable technological base posited by Article 7 of the TRIPS Agreement. These include: increasing the pool of trained workforce able to understand and assimilate technology and improving the quality of higher education institutions and scientific infrastructure, as well as creating networks between these educational and research institutions and

---

[11] Tobias Schmidt, 'Absorptive Capacity—One Size Fits All? A Firm-Level Analysis of Absorptive Capacity for Different Kinds of Knowledge' (2010) 31 MDE 1, 1; Zahra and George (n 7).

[12] David C Mowery and Joanne E Oxley, 'Inward Technology Transfer and Competitiveness: The Role of National Innovation Systems' (1995) 19 CambJEcon 67, 67–69; Wolfgang Keller, 'Absorptive Capacity: On the Creation and Acquisition of Technology in Development' (1996) 49 JDE 199, 200–02.

[13] Mowery and Oxley (n 12) 200–02.

[14] Dominique Foray, 'Technology Transfer in the TRIPS Age: The Need for New Types of Partnerships between the Least Developed and Most Advanced Economies' [2009] IPSDS 43–44.

[15] ibid.

enterprises, and better access to finance and efficient institutions.[16] Scholars have found that the capacity to innovate and to adapt to change depends on the science and technology system, the financial system, labour market institutions, and the education and training infrastructure.[17] Some have therefore argued that the interventions required to improve absorptive capacities must focus on the public agencies that support and/or perform R&D. These include universities, which perform research and train scientists and engineers; firms that invest in R&D and the exploitation of new technologies and public programmes intended to support technology adoption; and the laws and regulations on IP.[18]

Apart from the laws and regulations on IP, which have shown a positive trend, the other interventions are currently absent on the African continent,[19] thus hindering the acquisition and the absorption of technology. It is for this reason that some of those basic conditions will be subsequently explored; these include: government institutional frameworks focused on science, technology, and innovation; funding of R&D; development of human capital; and public policies to support technology adoption, such as the encouragement of the use of utility models to promote indigenization of foreign technologies, technology adaptation, and use.

a) Government institutional frameworks focused on science, technology, and innovation

It has been argued that, to achieve tangible results in promoting technological progress and innovation, there is a need to set up a dedicated institutional framework, namely a ministry responsible for science, technology, and innovation.[20] Governments in the fast-growing states of Asia—China, Indonesia, Japan, Korea, Malaysia, the Philippines, Taiwan, Thailand, and Vietnam— have endeavoured to establish separate ministries of science and technology to focus solely on science and technology and innovation issues.[21] It is therefore

---

[16] European Union, 'Report on the Implementation of Article 66.2 of the TRIPS Agreement' (WTO) Doc IP/C/R/TTI/EU/3 23–24 <https://docs.wto.org/dol2fe/Pages/FE_Search/FE_S_S009-DP.aspx?language=E&CatalogueIdList=288475,288461,288378,288403,288245,288236,288039,288037,288041,279272&CurrentCatalogueIdIndex=0&FullTextHash=&HasEnglishRecord=True&HasFrenchRecord=True&HasSpanishRecord=True> accessed 29 October 2023; Ana Pueyo and others, 'How to Increase Technology Transfers to Developing Countries: A Synthesis of the Evidence' (2012) 12 ClimPol 320, 333.

[17] In the same vein, see Bengt-Åke Lundvall and Rasmus Lema, 'Growth and Structural Change in Africa: Development Strategies for the Learning Economy' in Padmashree Gehl Sampath and Banji Oyelaran-Oyeyinka (eds), *Sustainable Industrialization in Africa: Toward a New Development Agenda* (Palgrave Macmillan UK 2016) 340.

[18] Mowery and Oxley (n 12) 80.

[19] Lundvall and Lema (n 17) 331.

[20] Kim (n 4) 316–18.

[21] Ministry of Science and Technology (China); Ministry of Education, Ministry of Research, Technology and Higher Education (Indonesia); Ministry of Culture, Sports, Science and Technology (Japan); Ministry of Science and ICT (Korea); Ministry of Energy, Science, Technology, Environment & Climate Change (Malaysia); Department of Science and Technology (Philippines); Ministry of Science and Technology (Taiwan); Ministry of Science and Technology (Thailand); and Ministry of Science and Technology (Vietnam).

evident that the impressive results of these fast-growing states in Asia are not accidental, but rather the result of a deliberate move by their respective governments to appropriately encourage science, technology, and innovation by providing for a dedicated body at government level.

The establishment of science and technology policies, and legal and institutional frameworks, has been attempted since African states attained independence.[22] For example, Ghana established the Ghana Research Council in 1959, Nigeria established the National Council for Scientific and Industrial Research in 1966, and Kenya established the National Council for Science and Technology in the late 1970s. A common characteristic of these institutions is that they were sectoral, hence they were pursuing technology policies of the specific ministry and department in which they were integrated. Consequently, these vertical institutional frameworks did not yield results because they were poorly coordinated and underfunded.[23] As a result, science and technology failed to contribute to the overall development of those states. An attempt to redress this anomaly was made by setting up horizontal governance structures, through the establishment of super-ministries of science and technology, or integrating science and technology in already established ministries, but this approach also failed to produce tangible results.[24] The reasons adduced for this failure are related to the lack of an institutional base for innovation and inadequate human capital.[25] Furthermore, lack of guidance and, in particular, of a clear vision and policies to direct science and technology towards promoting development were also highlighted. This is evidenced by the attempts aimed at correcting this weakness by conceptualizing policy as an instrument to advance economic growth and structural change.[26]

In view of these challenges a common position was taken at the Organisation of African Unity (OAU) in 1979 through the Monrovia Declaration, in which African governments committed to putting science and technology in the service of development.[27] This was further reinforced by the Lagos Plan of Action adopted

---

[22] UNECA, 'Assessing Regional Integration in Africa VII: Innovation, Competitiveness and Regional Integration' 85 <https://repository.uneca.org/handle/10855/23013> accessed 8 October 2017.

[23] ibid.

[24] ibid. Almost all states in Africa have Ministries of Science and Technology, and six include in the designation the term 'innovation' such as Ministry of Higher Learning, Science, Technology and Innovation (Angola); Ministry of Scientific Research and Innovation (Cameroon); Ministry of Higher Education, Research and Innovation (Chad); Ministry of Environment, Science, Technology and Innovation (Ghana); Ministry of Higher Education, Training and Innovation (Namibia); and Ministry of Science, Technology and Innovation (Uganda).

[25] Banji Oyelaran-Oyeyinka, *How Can Africa Benefit from Globalization?: Global Governance of Technology and Africa's Global Exclusion* (African Technology Policy Studies Network 2004) 11.

[26] UNECA (n 22) 86.

[27] Organisation of African Unity, 'Monrovia Declaration of Commitment of the Heads of State and Government, of the Organization of African Unity on Guidelines and Measures for National and Collective Self-Reliance in Social and Economic Development for the Establishment of a New International Economic Order' AHG/ST. 3 (XVI) Rev 1 <https://archives.au.int/handle/123456789/835> accessed 12 December 2019.

in Lagos, Nigeria, in 1980.[28] Nevertheless, most African states did not adopt any implementing policies at the national level until the 1990s. The policies later adopted were invariably set as national priorities, such as education and human resource development, agriculture, energy, health, the environment, industry, IP protection, and transport and communications.[29] Several policies also highlighted the importance of transfer and adaptation of knowledge and technology to promote technology progress at the national level. To implement these policies, most states established ministries responsible for science, technology, education, and research, while others created special bodies, such as industrial research institutes and national innovation and research councils.[30]

The transition from the OAU to the African Union (AU) brought further impetus to the use of science and technology to promote development, such as through the adoption of the Consolidated Plan of Action for Science, Technology and Innovation in 2005 and its transformation in 2014 into the Science, Technology and Innovation Strategy for Africa 2024 (STISA-2024) as the continental framework for accelerating Africa's transition to an innovation-led, knowledge-based economy within the overall framework of the AU Agenda 2063.[31] Within that framework, heads of state and government reiterated their commitment to mainstreaming science, technology, and innovation in Africa's developmental efforts.[32]

Notwithstanding all these initiatives and commitments at the highest level, science and technology are not yet contributing, as desired, to the development of Africa. It can be observed therefore that although the Southeast Asian formula of having separate ministries of science and technology is valid, it is not a panacea for the lack of technological progress. African states have similar institutional frameworks to those in Southeast Asia or even those in western states, yet technological development is not forthcoming. The question is: what went wrong and how can this be fixed?

Consistent with the methodology applied by this book, it is submitted that context matters. The ministries of science and technology cannot be a mere reproduction of foreign frameworks without deep introspection into their own contexts in order to understand the real needs of each state. For example, the United Nations Economic Commission for Africa (UNECA) observed that the priorities defined in the innovation policies seem to reflect a standardized approach that does not vary across states.[33] That is suspicious and worrying and may be symptomatic of

---

[28] UNECA (n 22) 86.
[29] ibid 98–102.
[30] ibid 86; Oyelaran-Oyeyinka (n 25) 11.
[31] UNECA (n 22) 93.
[32] African Union, 'Science, Technology and Innovation Strategy for Africa (STISA-2024)' <https://au.int/en/documents/20200625/science-technology-and-innovation-strategy-africa-2024> accessed 7 October 2017.
[33] UNECA (n 22) 86.

the fact that those policies are not a result of local assessment but are simply reproducing the trends followed in the region. This must be corrected. The ministries of science and technology in Africa must undertake a more targeted survey in their own states in order to assist their own government to appropriately define the priorities of development of the state that suit the needs and aspirations of its people and integrate innovation in that state. As suggested in Chapter 6, this can also be done in collaboration with other international partners and can take advantage of technical assistance programmes as provided for by Article 67 of the TRIPS Agreement.

Secondly, although the establishment of the Ministries of Science and Technology was a credible attempt to abandon the vertical governance that confined technology policies to specific sectors, it appears that the new set-up is not fit for purpose. Science and technology are cross-cutting and the horizontal governance of science and technology is not achievable through a ministry that is hierarchical and at the same level as other government ministries. Therefore, it is worth attempting new approaches. Accordingly, it is instructive to analyse the example of Singapore, where issues of science and technology are coordinated under the prime minister's office, as this provides a more coherent strategic overview and direction to policies and plans with a view to efficiently achieving the national strategic objectives.[34] It has been observed that African states aim to build innovation-led, knowledge-based economies.[35] However, even after three decades of existence of ministries of science and technology on the African continent, this has failed to achieve that objective. Therefore, it is time to elevate this important topic to a higher level of importance. The proposal to place the cross-cutting topic of science and technology at the office of the prime minister or president aims to force all sectors of government to incorporate innovative approaches in their strategies, policies, and plans and use science and technology to that end. This appears to be the best way to implement the 2014 African Union commitment by heads of state and government to use science, technology, and innovation as a tool for Africa's development.[36] For a continent that is desperately struggling to see innovation contribute to its development, this proposal is not misplaced.

In conclusion, the Southeast Asian formula to establish a dedicated institutional framework to fast-track the process of mainstreaming science and technology into development policies and strategies in Africa must be contextualized on the continent. A more robust and pervasive intervention designed to enable science and

---

[34] See <www.pmo.gov.sg/topics/science-and-technology>; the intricacies of management of the research and development culture in Singapore can be found in Goh Chor Boon, 'Creating a Research and Development Culture in Southeast Asia: Lessons from Singapore's Experience' (1998) 26 AJSS 49.

[35] African Union (n 32) 21.

[36] 'Decision on Strategy for Science Technology and Innovation in Africa 2024' Doc EX.CL/839(XXV) <https://archives.au.int/bitstream/handle/123456789/161/Assembly%20AU%20Dec%20520%20%28XXIII%29%20_E.pdf?sequence=1&isAllowed=y> accessed 12 June 2019.

technology to contribute to the development of Africa is required. Upgrading the science and technology position at government level is imperative.

b) Improvement of funding for research and development

Pursuing technological progress requires sufficient funding for R&D. Both scholars and policymakers invariably cite funding as one of the essential elements of R&D, and ultimately, of innovation.[37] The argument in favour of the government spending money to fund R&D is that the private sector is reluctant to invest in risky and long-term projects that are replete with uncertainties regarding commercial viability and return on their investment.[38] Accordingly, there is a concern, especially with regard to basic research, that it will not receive adequate attention from the private sector. Although there are some measures designed to mitigate the high costs of investment in R&D in the private sector, such as rewards through the IP system, tax incentives, and encouragement of research partnerships, government support is still a decisive factor.[39] In actual fact, a strong positive relationship between public funding of R&D and economic development has been established.[40] Furthermore, it has been found that government assistance can accelerate transfer of technologies from laboratories to industry.[41] Apart from the positive effect on skills development, and training and economic development in general, it is reported that government investment in R&D may pay off handsomely. This was evidenced by research on public funds utilized for agricultural innovation in Africa, which demonstrated that only 10 per cent of the projects failed while the remaining 90 per cent provided high, and in some cases even 'blockbuster', returns.[42]

The experience of Asian states shows that the fastest growing states are also those that allocate substantial funding to R&D in proportion to their GDP, as the following statistics illustrate: China (2 per cent), Korea (4.3 per cent), Malaysia (1.3 per cent), Singapore (2.2 per cent), and Thailand (0.5 per cent).[43] These investments in R&D and, correspondingly, in the development of human capital in

---

[37] Lundvall and Lema (n 17) 340–42; UNECA (n 22) 87; 'Decision of the Executive Council of the African Union at the Eighth Ordinary Session' (adopted 21 January 2006) Doc EX.CL/Dec.254 (VIII)).

[38] Rafael A Corredoira, Brent D Goldfarb, and Yuan Shi, 'Federal Funding and the Rate and Direction of Inventive Activity' (2018) 47 ResPol 1777, 1777; Richard Senter, 'The Impact of Government Research and Development Spending and Other Factors on State Economic Development' (1999) 23 PAQ 368, 368–84; Bronwyn H Hall and Josh Lerner, 'The Financing of R&D and Innovation' in Bronwyn H Hall and Nathan Rosenberg (eds), *Handbook of the Economics of Innovation*, vol 1 (North-Holland 2010) 611.

[39] Viviana Fernandez, 'The Finance of Innovation in Latin America' (2017) 53 IRFA 37, 37; Hall and Lerner (n 38) 611.

[40] Senter (n 38) 370.

[41] Geoffrey Hsu, 'The Competitive Edge: Approaches to Funding Scientific Research' (1995) 18 HarvIntRev 68, 85.

[42] William A Masters, 'Paying for Prosperity: How and Why to Invest in Agricultural Research and Development in Africa' (2005) 58 JIA SIPA 35, 56–59.

[43] UNESCO, 'How Much Does Your Country Invest in R&D?' <www.uis.unesco.org/_LAYOUTS/UNESCO/research-and-development-spending/index-en.html> accessed 9 February 2019.

Asian states are the backbone of the development of their technological absorptive capacities that are witnessed today.

Conversely, the funding of science and technology in Africa has always been challenging. This is so notwithstanding various government commitments to improve the current scenario, including the Decision of the Executive Council of the African Union that endorsed the call on member states to raise their national science and technology budget to 1 per cent of GDP and the encouragement to consider that as the benchmark in STISA-2024.[44] Nevertheless, African states hardly ever reach the threshold of 1 per cent spending on R&D in their respective states, despite the pledge by heads of state and government in the context of the African Union Summit. Indeed, it is reported that despite these commitments, only three states in Africa, namely Malawi, Uganda, and South Africa, spend more than 1 per cent of their GDP on research and development.[45] As it stands currently, R&D relies on donor funding and technical assistance or public disbursements.[46] However, this foreign assistance is shrinking and public spending on R&D is by no means satisfactory. This calls for innovative ways of securing funding, including from universities and R&D institutions themselves. That is the essence of the joint initiative between the World Intellectual Property Organization (WIPO) and the African Regional Intellectual Property Organization (ARIPO) for the development of guidelines on the elaboration of an IP policy and strategy for effective use of the IP system by universities and R&D institutions in Africa.[47] The aims of this initiative are to exploit the tremendous potential of African research outputs, ensure their ownership and protection through the IP system, promote their commercialization, and ultimately foster sustainability of Africa's R&D institutions with a view to better serving Africa's people.[48] It is to be observed, however, that these innovative ways of seeking university funding will never be enough to sustain R&D on the continent. Some have therefore proposed the establishment of institutions that could play the role of development banks in order to facilitate funding for R&D, drawing their finances from export levies on non-renewable commodities.[49] These

---

[44] 'Decision of the Executive Council of the African Union at the Eighth Ordinary Session' (adopted 21 January 2006) Doc EX.CL/Dec.254 (VIII)) (n 37) 41.

[45] See <www.resakss.org/node/6285>. This claim is not confirmed by the UNESCO Institute of Statistics, that indicates that South Africa's investment in research and development is at 0.8 per cent of its GDP, Uganda at 0.2 per cent, and no information is available for Malawi. The same source places Tunisia at 0.7 per cent, Egypt at 0.6 per cent, Ethiopia at 0.6 per cent, and Nigeria at 0.2 per cent. The South African Research and Development Strategy (2002) indicates that there was a drop in national research and development spending from 1.1 per cent in 1990 to 0.7 per cent of GDP in 1994 due to the end of key technology missions, such as military dominance on the continent and energy self-sufficiency.

[46] UNECA (n 22) 96.

[47] ARIPO and WIPO, 'Guidelines on Elaboration of Intellectual Property Policy and Strategy for an Effective Use of the IP System by Universities and Research and Development Institutions in Africa'.

[48] ibid.

[49] Lundvall and Lema (n 17) 341–42.

proposals remain on paper without any concrete operationalization and meanwhile the critical situation of lack of funding remains.

In conclusion, funding research must become a priority to enable Africa to benefit from the knowledge-based economy. However, it is clear that even if government spending is prioritized and the agreed continental benchmarks are met, it will not be enough to fund research, hence the need to devise creative mechanisms of funding, including collaboration with the private sector and empowering universities to exploit the outputs of their research. If those initiatives are undertaken, more funds may eventually be raised and more tangible results may be seen in the future.

c) Development of human capital to enable technological progress

Human capital has been regarded as one of the key factors in absorbing technology and promoting innovation.[50] Human capital has been defined as the total stock of knowledge, skills, competencies, and innovative abilities possessed by the population.[51] Studies undertaken in different parts of the world, such as India, Korea, Pakistan, Singapore, and Sudan, have widely demonstrated the correlation between human capital, technology, and economic growth.[52] It has specifically been argued that a skilled labour force has a huge impact in the production process of a state. This implies that an improvement in the quality of human capital results in an increase in economic growth.[53] Some studies have further demonstrated that

---

[50] Patrick van der Heiden and others, 'Necessitated Absorptive Capacity and Metaroutines in International Technology Transfer: A New Model' (2016) 41 JET-M 65, 73; Oyelaran-Oyeyinka (n 25) 14; Cassandra Mehlig Sweet and Dalibor Sacha Eterovic Maggio, 'Do Stronger Intellectual Property Rights Increase Innovation?' (2015) 66 WorldDev 665, 674.

[51] Theodore W Schultz, 'Investment in Human Capital' (1961) 51 AER 1, 1–17; OECD, *The Well-Being of Nations: The Role of Human and Social Capital* (OECD Publishing 2001) 17–19; David Ugal and Peter Betiang, 'Challenges for Developing Human Capital in Nigeria: Global-Local Connection' (2009) 4 <www.researchgate.net/publication/228313475_Challenges_for_Developing_Human_Capital_in_Nigeria_Global-Local_Connection> accessed 1 February 2019.

[52] Jimmy Alani, 'Effects of Technological Progress and Productivity on Economic Growth in Uganda' (2012) 1 PEFin 14, 14–23; Muhammad Amir, Bilal Mehmood, and Muhammad Shahid, 'Impact of Human Capital on Economic Growth with Emphasis on Intermediary Role of Technology: Time Series Evidence from Pakistan' (2012) 6 AJBM 280, 280–85; Khalafalla Ahmed Mohamed Arabi and Suliman Zakaria Suliman Abdalla, 'The Impact of Human Capital on Economic Growth: Empirical Evidence from Sudan' (2013) 4 ResWorldEcon 43, 43–53; Rajabrata Banerjee and Saikat Sinha Roy, 'Human Capital, Technological Progress and Trade: What Explains India's Long Run Growth?' (2014) 30 JAE 15, 15–31; Manuel Mendes de Oliveira, Maria Costa Santos, and BF Kiker, 'The Role of Human Capital and Technological Change in Overeducation' (2000) 19 EconEduRev 199, 199–206; Jong-Wha Lee, 'Human Capital and Productivity for Korea's Sustained Economic Growth' (2005) 16 JAE 663, 663–87.

[53] Chindo Sulaiman and others, 'Human Capital, Technology, and Economic Growth: Evidence From Nigeria' (2015) 5 SAGE Open 1–10 <https://journals.sagepub.com/home/sgo>; Monday Adawo, 'Has Education (Human Capital) Contributed to the Economic Growth of Nigeria?' (2011) 3 JEIF 46, 46–58; Badamasi Usman Babangida, 'Human Capital Development and Economic Growth in Nigeria' (2022) 2 Polac Management Review (PMR) 275; Ogujiuba Kanayo, 'The Impact of Human Capital Formation on Economic Growth in Nigeria' (2013) 4 J Econ 121–32; Aahad M Osman-Gani, 'Human Capital Development in Singapore: An Analysis of National Policy Perspectives' (2004) 6 ADHR 276, 279; Moses Ekperiware, sl Oladeji, and Olalekan Yinusa, 'Dynamics of Human Capital Formation and Economic Growth in Nigeria' (2017) 10 JGRESS 139, 140–43; Biswajit Maitra, 'Investment in Human

economic growth has mostly been influenced by the growth of the tertiary education sector. This is because expanding tertiary education may accelerate technological catch-up. For example, increasing the attainment of tertiary education by one year increases the rate of technological catch-up by 0.63 per cent a year, corresponding to 3.2 per cent in five years.[54] It has also been established that each additional year of education can yield 10 per cent wage increments; a one-year increase in average tertiary education levels would raise annual GDP growth by 0.39 per cent in Sub-Saharan Africa and increase African GDP per capita by 12 per cent in the long run.[55]

Some East Asian states stimulated their technological development through prioritizing the development of human capital as the basis for future technological and economic development.[56] One relevant case to be analysed in this regard is Singapore. Singapore acquired its independence in 1965, at the same time that African states were also achieving independence from the European colonial masters.[57] However, Singapore initially appeared to be doomed to failure because it lacked capital, natural resources, and land (consisting of one main island with a total surface area of 225 square miles and some fifty-three small islands) and had a population of only 1,860,651~~1,879,57~~, most of whom were neither highly skilled nor highly educated.[58] Yet, in a short space of time Singapore was able to reverse the situation and become one of the most developed states in Asia, even becoming the financial centre in the region and hosting the manufacturing operations and R&D activities of several multinationals.[59] This was possible because Singapore invested heavily in education, so much so that in the period 1991–95 it had the highest per capita national expenditure on education in Asia.[60] The state defined investment in human capital as its main priority.[61] However, the most striking feature of the Singapore experience is the fact that the strategy for the development of human capital was inextricably linked to the economic and national strategies for the development of the state.[62] In view of the fact that the challenge in linking

Capital and Economic Growth in Singapore' (2016) 17 GloBusRev 425, 425–37; David E Bloom and others, 'Higher Education and Economic Growth in Africa' (2014) 1 IJAHE 22, 122–42.

[54] Ekperiware, Oladeji, and Yinusa (n 53) 142.
[55] World Bank, *Accelerating Catch-up: Tertiary Education for Growth in Sub-Saharan Africa* (World Bank 2009) xxi <http://hdl.handle.net/10986/2589> accessed 15 January 2019.
[56] Larry E Westphal, 'Technology Strategies for Economic Development in a Fast Changing Global Economy' (2002) 11 EconInnovNewTechnol 275, 280; Oyelaran-Oyeyinka (n 25) 14.
[57] Osman-Gani (n 53) 284.
[58] 'Population Pyramids of the World from 1950 to 2100' (*PopulationPyramid.net*) <www.populationpyramid.net/ singapore/1965/> accessed 9 May 2023.
[59] Wong Tai Chee, 'Workforce Productivity Enhancement and Technological Upgrading in Singapore' (1997) 14 ASEAN EconBul 46, 55; Chong Li Choy, 'Singapore's Development: Harnessing the Multinationals' (1986) 8 CSEA 56, 56–69.
[60] Osman-Gani (n 53) 276–77; Chee (n 59) 47.
[61] Osman-Gani (n 53) 276.
[62] ibid 279; Maitra (n 53) 425–37.

national education to technological progress on the African continent is one of the main obstacles that the continent is facing it would be beneficial to understand how Singapore succeeded in this regard.

The first step that Singapore took was to articulate a clear definition of the state's human capital needs in the medium (three to five years) to long term (five to ten years) under the coordination of the ministry responsible for manpower.[63] Subsequently, it developed strategies to meet those needs, and this included the detailed determination of the numbers of different skills to be developed by universities, polytechnics, schools, and institutes of technical education. Therefore, it is apparent that the profile of the Singapore education system is flexible and is determined from time to time by the human capital needs of the state so as to define the human capital development strategies. In view of that, the universities have very limited places and instead there are filters that direct students into the relevant vocations and institutions.[64] The positive effect of the investment in education in the state's development is indisputable.[65] This is because skilled human capital is actively involved in the production process, increases productivity, and hence brings about growth in the income of the state.[66]

From this illustration it is submitted that the human capital strategy adopted by Singapore produced the required results because it was adequately mainstreamed into national strategies of the state, effectively contributing to Singapore's economic development. Therefore, the technological development of Singapore was not spurred primarily by technological transfer.[67] Instead, the state pursued its technology-intensive industrial development during the first phase by attracting foreign direct investment through investing in the development of human capital and infrastructure, which resulted in the state becoming a prime location for multinational companies.[68] This experience demonstrates that it is through the empowerment of human capital that the state will be able to attract investment that contributes the necessary technology for the state's development.[69] Thus, Singapore only moved to specifically stimulate the acquisition of technology or encourage investment in its creation around 1990 when technological capabilities were created in its manufacturing sector and has developed capacity to absorb foreign technology.[70]

---

[63] Osman-Gani (n 53) 281.
[64] ibid.
[65] Elena Pelinescu, 'The Impact of Human Capital on Economic Growth' (2015) 22 ProcediaEconFin 184, 184–85.
[66] Maitra (n 53) 432; Ekperiware, Oladeji, and Yinusa (n 53) 140.
[67] Westphal (n 56) 280.
[68] ibid; Ali M Nizamuddin, 'Multinational Corporations and Economic Development: The Lessons of Singapore' (2007) 82 IntSocSciJ 149, 156–57.
[69] Paul Romer, 'Human Capital and Growth: Theory and Evidence' (1990) 32 Carnegie-Rochester CSPP 251, 31.
[70] Westphal (n 56) 280.

African states should consider some elements of the Singapore experience that are relevant and practicable and adapt them to their specific contexts with a view to fast-tracking the much-desired technological progress. In this respect, the United Nations Educational, Scientific and Cultural Organization (UNESCO) has set global benchmarks with regard to financing education, positing that governments must allocate at least 4–6 per cent of GDP and/or at least 15–20 per cent of public (national) expenditure to education.[71] On average, Sub-Saharan Africa allocates 4.1 per cent of its GDP and 16.3–18.5 per cent of the national budget.[72] Sub-Saharan African states also allocate 20 per cent of their education budgets specifically to tertiary education.[73] These numbers would suggest that the region is on the right track, but the data regarding individual states demonstrates exactly the opposite, especially in Africa's LDCs. Indeed, the information available reveals clear signs of low expenditure in education that will affect their development and their readiness for technological catch-up. Many African LDCs are not achieving the recommended benchmark of allocating at least 4–6 per cent of their GDP to education as can be seen from the available data corresponding to 2017, which indicates that the Democratic Republic of Congo only contributed 1.5 per cent, Guinea 2.2 per cent, Liberia 3.8 per cent, Rwanda 3.2 per cent, South Sudan, 1 per cent, and Uganda 2.6 per cent, to name but a few.[74]

Even the stronger economies on the continent, such as Nigeria and South Africa, are grappling with challenges in financing education. In the case of Nigeria, the available statistics suggest that the expenditure in education is below the UNESCO recommended benchmarks. In fact, the percentages of the national budget allocated to education in a snapshot of a few selected years are as follows: 6.3 per cent in 2005, 7.8 per cent in 2006, 8.7 per cent in 2007, 6.42 per cent in 2009, 8.7 per cent in 2013, and 10.7 per cent in 2014, contrasted against the recommended 15–20 per cent.[75] This data shows that although a slight increase was recorded, it is still far below the recommended allocation.[76] It was also recorded that for 2018

---

[71] UNESCO, 'Education 2030: Incheon Declaration and Framework for Action: Towards Inclusive and Equitable Quality Education and Lifelong Learning for All' <http://uis.unesco.org/en/document/education-2030-incheon-declaration-towards-inclusive-equitable-quality-education-and> accessed 14 February 2019; UNESCO, 'Global Education Monitoring Report, 2019: Migration, Displacement and Education: Building Bridges, Not Walls' (UNESCO 2019) 235 <https://unesdoc.unesco.org/ark:/48223/pf0000265866> accessed 2 January 2020.
[72] UNESCO, 'Global Education Monitoring Report, 2019: Migration, Displacement and Education: Building Bridges, Not Walls' (n 71) 236; World Bank, *Financing Higher Education in Africa* (World Bank 2010) 210.
[73] World Bank, *Accelerating Catch-up* (n 55) xxvi.
[74] 'Government Expenditure on Education, Total (% of GDP)' (*World Bank Open Data*) <https://data.worldbank.org/indicator/SE.XPD.TOTL.GD.ZS?view = map&year = 2017&year_high_desc = false> accessed 7 February 2019.
[75] Lucas Elumah and Peter Shobayo, 'Effect of Expenditures on Education, Human Capital Development and Economic Growth in Nigeria' (2017) 3 NileJBE 41; Aramide Akintimehin, 'Effects of Human Capital Investment on Economic Growth in Nigeria' (2018) <https://www.researchgate.net/publication/325107498_EFFECTS_OF_HUMAN_CAPITAL_INVESTMENT_ON_ECONOMIC_GROWTH_IN_NIGERIA> accessed 6 February 2019.
[76] Elumah and Shobayo (n 75) 41.

the amount budgeted for education corresponds to 7.04 per cent of the total national budget, thus maintaining the trend whitnessed in the previous years.[77] There are accordingly sufficient elements to safely state that there is wholly inadequate funding for education in Nigeria.

The scenario in South Africa is heavily influenced by the negative legacy of the system of apartheid, although there is a trend indicating some improvement.[78] Indeed, the apartheid system of inferior 'Bantu education' was deliberately designed to restrict the productivity of black pupils in the national economy with a view to channelling them into subservient tasks and render them less competitive.[79] Today, South Africa is a free and democratic state with an upper-middle income economy that is based on mineral resources and manufacturing. The state needs to grow and be competitive in terms of productivity like its peers in the global industrializing economies, and the government is fully conscious that to achieve this, education is a key determinant.[80] Consequently, since the early 1990s South Africa has been allocating approximately 8 per cent of its GDP to education,[81] which is higher than the average of 4 per cent found in other upper-middle income states.[82] This has had a positive impact on the development of human capital with visible benefits for the economic development of the state and even the South African Development Community (SADC) region.[83]

This scenario suggests that all African states should consider the recommended thresholds set by UNESCO to finance education and undertake bold steps to achieve them in order to enable the improvement of the education systems and, consequently, the development of human capital on the continent. In fact, public investment in education cannot be neglected if the continent aspires to develop an innovation-led, knowledge-based economy.

---

[77] Azeezat Adedigba, 'FACT CHECK: Did UNESCO Ever Recommend 26 per Cent Budgetary Allocation to Education?' *Premium Times Nigeria* (9 December 2017) <www.premiumtimesng.com/news/headlines/251927-fact-check-unesco-ever-recommend-26-per-cent-budgetary-allocation-education.html> accessed 6 December 2017.

[78] Anthony Lemon and Lucy Stevens, 'Reshaping Education in the New South Africa' (1999) 84 Geography 222, 222–23.

[79] ibid 223; Edgar H Brookes, *Apartheid: A Documentary Study of Modern South Africa* (Routledge 2022) 57; Pierre-Carl Michaud and Désiré Vencatachellum, 'Human Capital Externalities in South Africa' (2003) 51 EDCC 603, 603.

[80] Government of South Africa, 'Growth, Employment and Redistribution Strategy' (GEAR 1996) <https://www.gov.za/sites/default/files/gcis_document/201409/gear0.pdf> accessed 30 October 2023.

[81] UNICEF, 'Education Budget—South Africa 2017/2018' <www.unicef.org/esa/media/2536/file/UNICEF-South-Africa-2018-2019-Education-Budget-Brief.pdf> accessed 2 June 2019.

[82] Johannes W Fedderke and John M Luiz, 'Does Human Capital Generate Social and Institutional Capital? Exploring Evidence from South African Time Series Data' (2008) 60 OEP 649, 665–68.

[83] Michaud and Vencatachellum (n 79) 604–05; Meenal Shrivastava and Sanjiv Shrivastava, 'Political Economy of Higher Education: Comparing South Africa to Trends in the World' (2014) 67 HighEduc 809, 809–22. It is also worth highlighting that the 1997 SADC Protocol on Higher Education and Training for the Nations of Southern Africa assigns the responsibility for developing human resources in the region to South Africa. Consequently, South African universities do not charge higher fees than equivalent domestic students to those originating from SADC states.

d) Adequate exploitation of science, technology, engineering, and mathematics (STEM)

The much publicized race to build knowledge-based economies in African states is certainly a laudable move, but it is submitted that this must be predicated on the strategic use of science and technology. This requires a major shift in focus towards including the teaching of STEM subjects in the educational systems in African states.[84] Ukeje synthesized this eloquently by saying:

> Without mathematics there is no science, without science there is no modern technology and without modern technology there is no modern society. In other words, mathematics is the precursor and the queen of science and technology and the indispensable single element in modern societal development. So, if any nation must develop, the study of science, technology and mathematics should be given adequate attention in the various levels of her education.[85]

For far too long, the relevance of this was not adequately considered and education favoured arts to the detriment of mathematics teaching in Africa. Indeed, it is maintained that the proportion of students that are enrolled in STEM in Africa averages less than 25 per cent.[86] This is even more evident if one applies the Harbison-Myers technical enrolment index. According to this system, Norway is ranked first with 73.52 per cent, South Africa, the most industrialized state in Sub-Saharan Africa, achieves only 23.61 per cent, Nigeria, 5.85 per cent, and in general most Sub-Saharan African states range from between 1 per cent and 5 per cent.[87] This is worrying as some have argued that the education system on the continent is 'unsuitable for industrialisation' and the failure to rebalance the huge enrolments in the humanities and social sciences towards STEM subjects, especially in tertiary education, is a serious threat to the sustainability of its economic development.[88] The current status in terms of enrolment in STEM in selected states is further analysed with a view to extracting some lessons for the African LDCs.

Nigeria has long recorded a continuous imbalance in its admission ratio between arts as compared with science, technology, and mathematics, in favour of arts.[89] Clearly, this has had a negative effect on the state's efforts to promote

---

[84] Subjects that fall under the term 'STEM' include, among others, natural sciences, such as biology, chemistry, and physics; and technology-related subjects, including computing and agriculture, technical drawing, and domestic science.

[85] Bennett Ujeke, 'The Challenges of Mathematics in Nigeria's Economic Goals of Vision 2010: Implication for Secondary School Mathematics', Paper presented at the 34th Annual National Conference of the Mathematical Association of Nigeria (MAN), 1–6 September 1997 293–96.

[86] Oyelaran-Oyeyinka (n 25) 20.

[87] ibid 14.

[88] Nkem Kumbah, 'STEM Education and African Development' 86 <http://africapolicyreview.com/stem-education-and-african-development/> accessed 7 February 2019; Oyelaran-Oyeyinka (n 25) 14.

[89] Lawrence I Aguele and Uche NV Agwagah, 'Female Participation in Science, Technology and Mathematics (STM) Education in Nigeria and National Development' (2007) 15 JSS 121, 121.

technological learning and had to be corrected through macro-level policies. That being the case, the federal government of Nigeria made important reforms in the National Education Policy in 1998 and 2004. On the one hand it was resolved that science and technology should be taught in an integrated manner in schools with a view to promoting students' appreciation of the practical application of basic ideas.[90] On the other hand, the policy demands that not less than 60 per cent of places shall be allocated to science and science-oriented courses in conventional universities and not less than 80 per cent in universities of technology.[91] The move to expressly establish a ratio of admission that clearly favours science, technology, and mathematics in Nigeria is a positive initiative and will certainly reposition science and technology in the national education system.

The government of Ethiopia has also established a ratio of 70 per cent of enrolments in tertiary education institutions in the natural and physical science and technology fields, and a residual 30 per cent in the social sciences and humanities.[92] Furthermore, the government introduced structured technical and vocational education and training and established two technical universities that focus on technology. As a result of these measures, the enrolment rate for technology courses quadrupled between 2008 and 2013.[93] In 1994, an Education and Training Policy was adopted and, as a result, the number of technical and vocational education and training institutions increased further.[94] These measures demonstrate once again that targeted policies, especially the imposition of a ratio between arts and science that favours the STEM subjects, can have a huge impact in the short term with regard to the number of enrolments. If this trend is consistent, an improvement in Ethiopia's technological capacity and ability to catch up in the medium to long term, can be expected.

Compared with Nigeria and Ethiopia, the case of South Africa is not very different, notwithstanding its status as one of the most industrialized states in Africa. During the period 2006–10, the proportion of STEM-based studies was 26 per cent in total, but with a tendency to shift more towards STEM studies at doctoral level, where STEM studies presently stand at 51 per cent.[95] Vocational education and training has been showing a declining trend. Whereas 13,500 artisans graduated in

---

[90] Nigeria Federal Government, *National Policy on Education* (4th edn, Nigeria Federal Government 2004) 35.
[91] ibid 39.
[92] African Capacity Building Foundation (ACBF), 'Africa Capacity Report 2017: Building Capacity in Science, Technology and Innovation for Africa's Transformation' (*Africa Portal*, 1 February 2016) 59 <www.africaportal.org/publications/africa-capacity-report-2017-building-capacity-in-science-technology-and-innovation-for-africas-transformation/> accessed 2 February 2019.
[93] ibid.
[94] ibid.
[95] Michael Kahn, *Science, Technology, Engineering and Mathematics (STEM) in South Africa* (Australian Council of Learned Academies 2013) 16–17; Kelly Roberts, 'Securing Australia's Future STEM: Country Comparisons' (Australian Council of Learned Academies) <https://acola.org/wp-content/uploads/2018/12/Consultant-Report-South-Africa.pdf> accessed 4 February 2019.

1985, this number shrank to only 2,548 by 2004.[96] As a middle-income economy with a strong manufacturing sector, South Africa is striving to correct this situation and develop STEM and information and communication technology knowledge and skills in order to promote economic development and technological progress through several policies.[97] For example, the Revised National Strategy for Mathematics, Science and Technology Education in General Education and Training and Further Education Training (2019–2030) prioritizes the following targets: 450,000 learners qualifying to study towards mathematics and physical sciences qualifications by 2030; 30,000 artisans produced per year; learners achieving more than 50 per cent in mathematics; increased participation rates in technical and vocational training colleges to 25 per cent; and 90 per cent of Grades 3, 6, and 9 obtaining more than 50 per cent in the national assessment.[98] It is expected that, once these targets are achieved, the current scenario will change and a more promising future for technology and innovation will prevail.

With regard to Kenya, notwithstanding the significant strides that the state has taken in terms of industrial development, it seems to be at odds with the examples provided. Indeed, it is observed that despite an increase in the demand for scientific expertise in industry, the number of students enrolling in STEM is declining.[99] This is a matter of great concern that can threaten the sustainability of that progress and needs to be addressed by providing mechanisms to attract more students enrolling in STEM and ensuring that they complete their degrees in order to feed the industry. Furthermore, the establishment of a ratio of enrolment in universities that favours STEM would be desirable in order to stop the declining STEM-related enrolments witnessed in the state. Regrettably, this scenario is the norm in the majority of African states and needs to be adequately tackled in order to change the current trend. Africa's LDCs need to view science and technology with a new and vigorous approach and strive to promote skills development through technical and vocational education and training and through science and technology universities, and further apply a ratio between arts and science that favours STEM. The examples of Ethiopia, Nigeria, and South Africa make it clear that in order for

---

[96] Kahn (n 95) 19.

[97] Department of Basic Education, South Africa, 'Action Plan to 2019 Towards the Realisation of Schooling 2030' <www.education.gov.za/Portals/0/Documents/Publications/Action%20Plan%202019.pdf?ver=2015-11-11-162424-417> accessed 5 May 2023; Department of Basic Education, South Africa, 'National Strategy for Mathematics, Science and Technology Education in General and Further Education and Training' <www.voced.edu.au/content/ngv%3A3777> accessed 2 June 2019; Vicki Gardner and others, 'Approaches to Strengthening Secondary STEM & ICT Education in Sub-Saharan Africa' (2018) Working Paper No 10 <www.bristol.ac.uk/media-library/sites/education/documents/Binder1.pdf> accessed 2 June 2019.

[98] Department of Basic Education, South Africa, 'National Strategy for Mathematics, Science and Technology Education in General and Further Education and Training' (n 97).

[99] Government of Kenya, 'Sector Plan for Science, Technology and Innovation 2013–2017' <www.planning.go.ke/wp-content/uploads/2020/11/SECTOR-PLAN-FOR-SCIENCE-TECHNOLOGY-AND-INNOVATION-2013-2017-1.pdf> accessed 6 February 2019.

Africa to be enabled to receive, absorb, and adapt technology for its own needs, the current science education must be restructured to adequately integrate and promote science, technology, and mathematics.[100] There is no doubt that science education is the only tool that can adequately prepare Africa for the challenges of the future and propel the continent into the knowledge-based economy.

e) Socialization or contextualization of science and technology

As highlighted at the beginning of Section B on 'The Development of Technology Absorption and Adaptation Capabilities as Enablers of Technology Transfer in LDCs', an emphasis on science and technology is synonymous with western culture, and Africa has been struggling to adapt to this phenomenon. To overcome this challenge and enable Africa to take advantage of the outputs of science and technology it has been proposed that a 'conceptual ecocultural paradigm' be adopted.[101] This paradigm will prompt the design of science education focusing on growth and development of the individual's perception of knowledge, based on the socio-cultural environment in which the learner lives and operates.[102] Indeed, science and technology is part of the social system, hence it is heavily influenced by societal factors such as culture, values, and beliefs.[103] Accordingly, it is crucial that the relationship between science, technology, and African society be contextualized so that it can substantively address the challenges the continent is facing. With that approach, the African environment, indigenous scientific and technological principles, theories and concepts, and values of typically African human feelings will gain prominence in shaping technology and this will serve to better the continent.[104] This will also enable each state to move away from a 'one size fits all' orientation and design policies that will take into account its own specific context, including cultural aspects, together with political and economic factors.[105]

As rightly highlighted by the Nigerian National Policy on Education, in the process of formulating policy on education there is a need to first identify the overall philosophy and goals of the state.[106] In light of that assertion, section 1 of the Policy is titled 'Philosophy and Goals of Education in Nigeria'. This is where the timely and most relevant idea of the 'conceptual ecocultural paradigm' should be integrated. The Nigerian Policy attempts to include philosophical pillars that lean towards this paradigm, such as 'the full integration of the individual into the community' and

---

[100] Thomas Young, Jonathan Cole, and Denice Denton, 'Improving Technological Literacy' (2002) 18 IssuesSciTechnol 77–79.
[101] Olugbemiro J Jegede, 'School Science and the Development of Scientific Culture: A Review of Contemporary Science Education in Africa' (1997) 19 IJSS 1, 16.
[102] ibid.
[103] Trust Saidi and Tania S Douglas, 'Towards the Socialization of Science, Technology and Innovation for African Development' (2018) 10 AJSTID 110, 110–13.
[104] Jegede (n 101) 16.
[105] Saidi and Douglas (n 103) 111.
[106] Nigeria Federal Government (n 90) 6.

strategic objectives of education that include 'training of the mind in the understanding of the world' and 'the acquisition of appropriate skills and development of mental, physical and social abilities and competencies as equipment for the individual to live in and contribute to the development of the society'.[107] However, in consideration of its relevance to the transformation of education in Africa and to enable a more contextualized technological development, it is submitted that the contents of the 'conceptual ecocultural paradigm' should be more evident by incorporating unambiguous concrete actions that operationalize the paradigm.

The African Union Science, Technology and Innovation Strategy for Africa 2024 (STISA-2024), which was designed to transform Africa through science, technology, and innovation implicitly embraces the contents of the paradigm.[108] STISA-2024 emphasizes that science, technology, and innovation programmes must take into account the views of the relevant stakeholders in research and innovation from both the public and private sectors as well as international development and funding partners. Only in this way will Africa be in a position to use scientific culture in everyday life and propel itself to technological progress and innovation. It is therefore expected that the policies to be adopted by African governments in this regard will be informed by the pressing call in STISA-2024 for the socialization of science, technology, and innovation policies in Africa. Indeed, it is necessary that the future policies of education take into account the local contexts in order for them to be relevant and contribute to technological progress based on local needs and local solutions. This is because technology can only succeed in the recipient state if it addresses the needs encountered at the local level. Therefore, technology must be contextualized to have the desired impact. The socialization of science and technology is thereby fully substantiated.

f) Technological development through acquisition and indigenization of foreign technologies

i) Utility models as a tool for indigenization of foreign technologies

There is no single or widely accepted legal definition of the concept 'utility model' within the IP paradigm.[109] Although the Paris Convention for the Protection of Industrial Property of 1883 includes, in Article 1(2), utility models among the categories of IP rights that should be protected by its members, it does not provide any definition. Utility models do not even feature as IP rights included in the

---

[107] ibid 7–8.
[108] African Union (n 32).
[109] Uma Suthersanen, 'Utility Models and Innovation in Developing Countries' (2006) <www.academia.edu/27254157/Utility_Models_and_Innovation_in_Developing_Countries> accessed 9 February 2019. The definition of utility models set by Andean Community Decision 486 of 2000 is illustrative: a 'new form, configuration, disposition of elements, of any artefact, tool, instrument, mechanism or other object or any part of the same, that permits a better or different functioning, use or manufacture of the object which incorporates or which offers any use, advantage or technical effect that it did not have previously'.

TRIPS Agreement as a category to be protected.[110] Furthermore, the concept itself is not universally used, as evidenced by the existence of terms such as 'petty patent', 'innovation patent' (Australia), 'utility innovation' (Malaysia), 'utility certificate' (France), and 'short-term patent' (Belgium) to describe the same content. Therefore, utility models can only be understood by their common features, namely the fact that they are a patent-like but second-tier patent system that offers a more economical and rapid protection to technical inventions, which would not qualify in normal circumstances to be protected through a fully-fledged patent mechanism.[111]

The utility models appear to be the most appropriate means to protect minor or incremental innovations. Usually they are effective in the protection of the functional aspects of a product and are very common in the mechanical, optical, and electronic fields.[112] In view of these characteristics, some have argued that they could serve as a useful tool for protecting the simple innovations developed by artisans, single innovators, and small and medium-sized enterprises in developing states.[113] The fact that utility models do not go through a process of examination, require only industrial applicability, and must exhibit a novelty that is limited to the territory concerned further militates in favour of the tool in developing states.[114]

Historically, utility models seem to have played a role in the technological catch-up of major industrial states such as China, Germany, India, Japan, and South Korea and are currently very popular in the Asian states.[115] Following these findings, it appears evident that utility models may be a game-changer with regard to technological catch-up of African states. However, it cannot be ignored that utility models have also attracted some criticism. One of the main criticisms of the utility models system is that it is susceptible to abuses that may hinder the achievement of its desired objectives. Utility models should, therefore, be viewed as a mechanism capable of fostering innovation especially by local innovators and small and medium-sized enterprises. Cognizance must, however, be taken of the fact that if caution is not exercised, the system may better serve the interests of big companies that can file utility models with the objective of ring-fencing their technologies against small innovations by local competitors. Furthermore, it must be acknowledged that strict enforcement by big companies may also lead to infringement proceedings whenever local innovators engage in small improvements of foreign technologies.[116] Although these argument are valid and well-grounded, it does

---

[110] Uma Suthersanen, Graham Dutfield, and Kit Boey Chow (eds), *Innovation Without Patents: Harnessing the Creative Spirit in a Diverse World* (Edward Elgar 2007) 20–21.
[111] ibid ix.
[112] ibid vii.
[113] ibid.
[114] ibid 1–2.
[115] ibid.
[116] Nishantha Sampath Punchi Hewage, *Promoting a Second-Tier Protection Regime for Innovation of Small and Medium-Sized Enterprises in South Asia: The Case of Sri Lanka* (1st edn, Nomos Verlagsgesellschaft mbH 2015) 261.

not change the importance of the utility model regime to developing states. What this argument instructs us to ensure is that the regime is accompanied by adequate measures to prevent abuses and still deliver the desired results to the target group, that is, local innovators and local small and medium-sized enterprises. As seen in the Asian cases, the utility models regime was developed and implemented with the primary objective of benefitting local innovators, hence these were granted mainly to nationals. Legally, this is possible because utility models were not included in the TRIPS Agreement, thus giving room to the WTO member states to freely regulate them without any risk of violating minimum standards.[117]

Another objection is that utility models may constitute a further limit on the freedom of copying and imitation as well as an erosion of the concept of public domain that serve to assist small and medium-sized enterprises.[118] The flaws in this argument lie in the fact that IP rights are viewed not as simple monopolies but as tools that are designed to provide incentives for innovation rather than restricting it. Hence, a well-designed utility models regime will strive to assist small and medium-sized enterprises and local innovators in viewing this IP right as a tool to foster innovation and not to impede it. Fears about the possible erosion of the public domain also do not stand because it will not be through the creation of new improvements that such a phenomenon will arise. Right on the contrary, the public domain will be enriched with the new solutions created by local innovators in order to better cater for local needs.

A further argument suggests that the proliferation of trivial patents will bring too many property rights into the system and the fact that these are not even filtered by a substantive examination will create uncertainty and lessen their real value. Here, the risk of evergreening and excessive litigation is also highlighted.[119] Although this argument is valid, it should be borne in mind that the utility model is an IP right that has been in existence for a long time (indeed, the German regime was introduced as early as 1891) and the main objective in the first stages of technological development is to facilitate technological learning and absorption. The proliferation of improvements may be a welcome signal that technology is being absorbed and adapted to local needs. In addition, utility models applications and grants provide property rights to local innovators and develop the culture of patents in the state that will constitute the bedrock for future technological progress. This will provide more certainty to local innovators who will find a simple way to secure ownership of their endeavours.

Finally, it was noted that some Asian states, such as South Korea and Taiwan, were forced to strengthen their IP rights regimes in view of the pressure that the United States (US) consistently placed on them under the Watch List of the Special

---

[117] Suthersanen (n 109) 3.
[118] Hewage (n 116) 261.
[119] ibid.

301 Report on Intellectual Property Rights.[120] It was argued that the flexibilities that assisted Asian states when they adopted weak IP regimes and promoted their technological development may not be available today due to the strengthening of IP systems and the minimum standards imposed by the TRIPS Agreement.[121] It is however prudent to seriously assess this concern in light of the fact that almost all LDCs adhere to international IP legal instruments that may constrain them in their freedom to customize their IP rights regimes. As it stands now, the advantages of the utility models seem to outweigh the risks that were expressed. Moreover, those risks may be mitigated or controlled with appropriate measures that must be crafted from the inception of the system. Furthermore, it is also necessary to state that the TRIPS Agreement has not imposed any minimum standards with regard to utility models, hence ample space for flexibility is available in this domain and must be exploited by African LDCs while it is still in existence.

The proposal to promote utility models in African states, especially LDCs, therefore stands firm. From the point of view of the legal framework, this seems to be the case because currently at least thirty African states provide for utility models protection and this includes all seventeen states of the Organisation africaine de la propriété intellectuelle (OAPI) (through Annex II on Utility Models of the Bangui Agreement 1999) and seven ARIPO member states through their national laws.[122] It is also worth highlighting that the ARIPO Harare Protocol on Patents and Industrial Designs of 1982 provides protection for utility models in twenty member states.[123] However, the reality on the ground shows that the system is yet to be exploited by local innovators, because Africa's share of global utility models applications accounts for less than 0.1 per cent.[124] Lack of protection in African states is also apparent in the patent system. Statistics from the WIPO Intellectual Property Indicators which have been consistent throughout the years show that patent applications filed in African IP offices account for less than 0.6 per cent, with residents accounting for only 17 per cent.[125] This results in more than 99 per cent of inventions falling within the non-protection layer. For this reason, the non-protection of technological outputs makes a strong case for the introduction and promotion of utility models as a seamless way of developing a patenting culture on the continent by way of the use of a less stringent mechanism. Although Africa stands to benefit from the utility models system to promote technological learning

---

[120] Kumar (n 4) 215.
[121] ibid.
[122] Isaac Rutenberg and Lillian Makanga, 'Utility Model Protection in Kenya: The Case for Substantive Examination' [2016] AJIC 19, 19–37; Adams & Adams, *Adams & Adams Practical Guide to Intellectual Property in Africa* (PULP 2012).
[123] 'Harare Protocol on Patents and Industrial Designs' <www.aripo.org/wp-content/uploads/2018/11/Harare-Protocol-2019.pdf> accessed 12 February 2019.
[124] WIPO, 'World Intellectual Property Indicators 2018' (2018) 84–86 <www.wipo.int/publications/en/details.jsp?id=4369> accessed 12 February 2019.
[125] ibid 39.

and improve its absorptive capacities, the continent is not yet taking advantage of it and this must be addressed. This requires that awareness creation campaigns be undertaken among innovators; incentives be provided for promoting the use of utility models by local innovators, such as exemptions or reductions in fee payments; support be offered in the commercialization of innovations; and awards for the best utility models established. Likewise, precise and focused policies and legislation must be introduced to promote this form of IP rights and it must be streamlined with national development policies.

ii) *The acquisition and indigenization of foreign technologies through utility models in East Asian states*

South Korea and Taiwan, the most technologically advanced East Asian states, invested in the acquisition of technology that was progressively indigenized to the local context by domestic enterprises through utility models and industrial designs.[126] The mechanism of protection of utility models established in South Korea, for example, favoured its own nationals, as evidenced by the fact that the country awards 92–95 per cent of the utility models and industrial designs to its own nationals.[127] It is claimed that the rapid technological growth recorded in South Korea from 1960 to 1980 was heavily facilitated by the duplicative imitation rooted in IP rights that South Korean enterprises could enjoy.[128] South Korea only changed this approach as a result of pressure from the US that has continuously placed the state on its Priority Watch List under the Special 301 Report on Intellectual Property Rights.[129] In February 1993, South Korea undertook to strengthen its IP rights protection and enforcement. This was formalized through the new Patent Act of 1995.[130] From the foregoing, the role of the South Korean authorities is evident, as they deliberately enacted laws and regulations and undertook initiatives that facilitated reverse engineering in order to provide an opportunity for local companies to gradually emerge as innovators. Thus, from the illustration of the South Korean model that was emulated by many other states it appears clear that the initial phases of its technological development relied primarily on technology transfer, by using technologies transferred from abroad through capital goods imports, turnkey plant construction, overseas training and technical education, and various forms of technical assistance.[131] However, this was phased out gradually

---

[126] Westphal (n 56) 281–82; Hulya Ulku, 'Technological Capability, Innovation, and Productivity in Least-Developed and Developing Countries' in Augusto López-Claros (ed), *The Innovation for Development Report 2010–2011: Innovation as a Driver of Productivity and Economic Growth* (Palgrave Macmillan UK 2011) 144; World Bank, 'Global Economic Prospects 2008' (n 5) 145; Kumar (n 4) 214–15.
[127] Kumar (n 4) 214–15.
[128] ibid 215.
[129] The 'Special 301 Report on IP Rights' is published by the Office of the United States Trade Representative, see <https://ustr.gov/issue-areas/intellectual-property/Special-301>.
[130] Korea Act 5080, 29 December 1995.
[131] Kim (n 4) 315.

as technological capabilities were being built at the national level and conditions were met for the development of home-grown inventions.[132] Consequently, 'invention policies' were adopted aiming at the acquisition of capabilities to handle more complex innovations and to use those capabilities to develop sophisticated innovations locally.[133] This experience suggests therefore that the quest for transfer of technology by LDCs is a good starting point to establish the basic conditions for future innovation but is not enough on its own. There is a need to undertake other measures that will facilitate technology absorption and the development of home-grown technologies, including through the promotion of utility models.

The same trend was seen in Japan in its early stages of development. The objective of patent protection was to contribute to the 'development of industry' and not as an end in itself.[134] During the period 1888–1975, there was no patent protection for food, beverages, and pharmaceutical products in Japan in order to enable local innovation to thrive in those areas.[135] Japan also made extensive use of utility models, since the introduction of utility models legislation in 1905, by protecting adaptations or improvements over the imported technology by domestic inventors.[136] For example, the Japan Patent Office deliberately took twenty-nine years to grant a patent filed by Texas Instruments in 1960. The patent was finally granted when the state had absorbed the technology, improved upon it, and controlled 80 per cent of the US market for computer semiconductors.[137] Utility models and industrial designs were deliberately used by the Japanese government to protect small improvements in original foreign inventions that were effected by local inventors.[138] Statistics show, for example, that in 1980, the Japanese Intellectual Property Office awarded 49,000 utility models to its nationals, as against 533 granted to foreigners.[139] A similar pattern was also observed in relation to industrial designs, with 31,000 being granted to Japanese nationals as against only 600 to foreigners.[140] The Japanese experience further suggests that there was a deliberate effort to promote more usage of utility models instead of patents in the first phase of technological development, resulting in a higher number of utility models applications during the period between 1905 and 1980.[141] That trend was only reversed in 1980 when, for the last time, Japan recorded 191,785 applications for utility models contrasted against 191,020 patent applications.[142] Interestingly,

---

[132] Westphal (n 56) 282.
[133] ibid.
[134] Suthersanen, Dutfield, and Chow (n 110) 142–45.
[135] Kumar (n 4) 214.
[136] ibid.
[137] Suthersanen (n 109) 142–45.
[138] ibid.
[139] ibid.
[140] ibid.
[141] KS Kardam, 'Utility Model: A Tool for Economic and Technological Development: A Case Study of Japan' (Tokyo Institute of Technology 2007) 52–56 <www.ipindia.gov.in/writereaddata/images/pdf/FinalReport_April2007.pdf> accessed 23 January 2019.
[142] ibid 53.

that extended period coincides with the most impressive phase of technological progress witnessed in Japan. During that period, the Japanese innovation system was still in its infancy and the patent system was unaffordable to local innovators. Therefore, those innovators made extensive use of the utility models system to protect the petty and small modifications of the imported technologies in order to suit their local needs. It is for this reason that it has been asserted that utility models protection was one of the factors that contributed to the economic and technological development of Japan.[143] Japan then reversed this strategy when it realized that its technology was sufficiently advanced to fill the gap between Japanese and western technologies and the number of patent applications had surpassed that of the utility models.[144]

Higher numbers of utility models in comparison with patents were also seen in China until 2003.[145] Indonesia, Malaysia, Philippines, Taiwan, Thailand, and Vietnam also followed the same trend.[146] In general, the model followed by the Asian states postulates the introduction of public policies and corporate strategies designed to build technological capabilities in the initial stages with a view to encouraging imitative reverse engineering of the acquired mature foreign products without infringing on IP rights.[147] In particular, the model hinges on the use of utility models as an effective way to fast-track technology learning and to build technological capabilities that will further foster innovation.

The use of the utility models system to promote innovation by local innovators has been the norm in most Asian states and has been highly successful.[148] It is accordingly submitted that the massive use of utility models in the initial phases of development facilitated the learning of technology and established the basic conditions to promote creativity and spur innovation in those Asian states. The greater number of patents currently filed in those states are reflective of their advanced stage of technological development that can now stand on a stronger patent culture, with more local enterprises seeking stronger protection for their own innovations.[149]

iii) Indigenization of foreign technologies through the use of utility models in Africa
The successful model applied in the Asian states appears to be a good recipe to be considered in Africa's LDCs. Indeed, the low level of inventiveness of many inventions in Africa prevents them from passing the strict patentability criteria of the patent system. The utility models therefore appear to be the ideal legal tool to

---

[143] ibid 68.
[144] ibid 43.
[145] ibid 91–92.
[146] ibid 102.
[147] Kim (n 4) 311–24; World Bank, 'Global Economic Prospects 2008' (n 5) 146–47.
[148] Kardam (n 141) 102.
[149] Kumar (n 4) 216.

protect those small but useful inventions that can promote local innovation. This is not yet the case in Africa where utility models are yet to attract much attention and are seldom used for the benefit of their innovators. This is clearly illustrated by the small number of utility models lodged in Africa, and African LDCs in particular. The World Intellectual Property Indicators 2018 reveal that 1,761,200 new utility model applications were filed worldwide in 2017, with Africa's share being a mere 192 applications, corresponding to less than 0.1 per cent.[150] It is also telling that the total number of applications filed since the introduction of utility models in the ARIPO system in 2001 stands at only 139, mainly originating from Zimbabwe, Kenya, and South Africa.[151] In these individual states, the number of utility model applications is not impressive. From 2000 to 2015, Kenya recorded only 466 applications; in the period 2000-10, Ethiopia only recorded 259 applications. Statistics regarding a single year are even more disappointing and reveal the extremely low level of applications that are lodged at the national level. For example, in 2017 the following numbers of utility models applications were filed nationally: Botswana, five; The Gambia, one; Ghana, seven; Kenya, 136; Mozambique, eight; Rwanda, nine.[152] In 2018, the figures were not very different: Botswana, twelve; The Gambia, zero; Ghana, two; Kenya, 178; Mozambique, nine; Rwanda, six; Tanzania, three; Uganda, thirteen.[153] The total number of utility models filed through the ARIPO system is 306, covering eighteen member states.[154]

The data analysed provides clear evidence to infer that, notwithstanding the tremendous benefits that Africa could derive from the use of utility models, the continent is not yet taking advantage of it. There are several factors impeding the use of IP that are of specific relevance to Africa. These include lack of awareness of the important role that utility models can play to fast-track technological learning and promote local innovation; a negative perception of utility models as a lesser route for the protection of innovative outputs, hence the focus on using patents as a more secure and prestigious mechanism; lack of skills and capacity to draft utility models specifications; lack of a clear policy from IP offices and governments to promote the utilization of utility models; and the high costs that are still associated with the utility models protection system.[155] As a result of this observation, it appears necessary that a clear strategy be devised at the national level to tackle these

---

[150] WIPO (n 124) 84–86. It is however highlighted that data was only collected from ARIPO and five national IP offices, namely Botswana, The Gambia, Ghana, Kenya, and Rwanda.
[151] ARIPO, 'Report of the Administrative Council (1999)' Doc ARIPO/AC/XXIII/19; ARIPO, 'Status of Operations of Industrial Property Rights' Doc ARIPO/AC/XLII/3.
[152] Rose Adhiambo Mboya, 'Utilization of Industrial Designs and Utility Models in Africa: A Case Study of Kenya' in Antony Taubman and others (eds) WIPO-WTO Colloquium Papers: 2018 Africa Edition (Research Papers from the 2018 Regional WIPO-WTO Colloquium for IP Teachers and Scholars in Africa) (2021) 139–54.
[153] ARIPO, 'ARIPO Annual Report 2018' 56.
[154] ibid.
[155] Mark Ndumia Ndungu, 'Promotion and Protection of Utility Models for Economic Development in Kenya' (LLM thesis, Africa University 2015).

challenges if African states genuinely aspire to catch up technologically. Box 7.1 contains some of the strategies that could be adopted to promote the use of utility models at the national level.

A practical example can illustrate this assertion further. The fields of technology in which most of the utility models applications were filed during the period 2000–10 in Ethiopia were agriculture (27 per cent) and chemicals (20 per cent).[156] These two areas correspond to the priority areas of strategic development of the state that include the eradication of extreme poverty and hunger, and the improvement of health standards and ensuring environmental sustainability.[157] It is therefore submitted that, if properly employed, utility models can potentially enable Ethiopian innovators to develop the required technologies in those priority areas with a view to assisting the state to overcome existing technological challenges. This is achievable if government efforts are channelled towards setting up appropriate strategies and policies that will persuade more innovators to direct their innovative endeavours towards the priority areas of development. The sample strategy proposed in Box 7.1 could guide government action to make appropriate use of utility models for the benefit of the state.

In conclusion, Africa needs to shift to utility models to fast-track the development of its absorptive capacities and technological learning, and to foster local innovation. Efforts to fulfil that aspiration through patents are misguided and will unnecessarily over-extend the weak African IP systems and will not have the desired results. Utility models appear to be the most effective IP right at this stage of development of African states, and in particular LDCs, to fast-track technological absorption and promote innovation.

g) Leveraging the digital economy to fast-track Africa's innovative capabilities

The technological revolution witnessed in the last decades, especially through the digital transformation, provided more hope for marginalized countries to promote development that is anchored in technological innovation.[158]

The digital transformation seems to offer to the developing economies an unprecedented opportunity to leapfrog many stages of technological progress and achieve rapid industrialization by leveraging the digital technologies. Indeed, these

---

[156] Abdulrazak Oumer Jeju and Kyomugasho Mercy Kentaro, 'Does the Utility Model Protection Regime Make Sense in Least Developing Countries? A Comparative Study of the Ethiopian and the Ugandan Utility Model Protection Regimes and Its Effect on Innovation and Economic Growth' (Africa University 2012) 23–24.

[157] Jeju and Kentaro (n 156); FAO, 'National Strategy and Action Plan for the Implementation of the Great Green Wall Initiative in Ethiopia' <www.fao.org/faolex/results/details/en/c/LEX-FAOC169534/> accessed 12 January 2018.

[158] Xiaolan Fu, Giacomo Zanello, and George Owusu, 'Innovation in Low Income Countries: A Survey Report' (*GOV.UK*, 2014) 3 <www.gov.uk/research-for-development-outputs/innovation-in-low-income-countries-a-survey-report> accessed 12 June 2019.

Box 7.1 National Strategy to Promote the Use of Utility Models

Under the existing IP systems, the protection of inventions is secured through the patent system. However, this system imposes very stringent requirements, such as global novelty, strong inventiveness criteria, and highly assessed industrial applicability. In addition, the procedure for the grant of patents is complex, long, and expensive. The simple inventions developed by local innovators in Africa do not satisfy the rigorous patentability criteria and fail to be captured, protected, and locally exploited to promote local innovation.

This calls for the establishment and promotion of a less demanding, second-tier form of protection through utility models. The utility models require only two criteria to be considered: industrial applicability and novelty, the latter being assessed only with regard to the national state-of-the-art in technology. Furthermore, utility models may be crafted at the national level to legally secure rights in a simple, rapid, and inexpensive way.

The following strategies will be adopted at the national level to promote the use of utility models:

1. Institutional framework

Ministries and other institutions dealing with the promotion of science, technology, technology transfer, innovation, and industrial development will be charged with the responsibility of promoting the use of utility models with a view to facilitating technology absorption and adaptation of foreign technologies to respond to local needs and foster local innovation through extensive use of minor and incremental innovations.

2. Legal framework

Caution will be exercised to ensure that the legal framework governing science, technology, technology transfer, innovation, and industrial development consistently includes the use of utility models as the most suitable way of securing IP rights on minor and incremental innovations.

3. Grant and administration of utility models

The system of administration of IP rights shall be conceived in such a way that the grant of utility models will benefit local single innovators, artisans, and small and medium-sized enterprises, and the procedures for securing the rights will be simple, rapid, and inexpensive. Unnecessary barriers shall be removed in order to ensure that local innovators make full use of the system. Accordingly, the national offices charged with the responsibility for granting IP rights shall remove the substantive examination of utility models or endeavour to make it less stringent. In particular, the novelty criteria shall only be assessed with regard to the national state-of-the-art, and the office shall be satisfied once the invention presents an unequivocal contribution to improving the functionality of previous inventions or causes it to better solve local problems.

4. Turning innovations into utility models

Specific support programmes shall be established to assist local innovators to turn their ideas into utility model applications. Innovation and technology transfer offices of academic and research institutions and IP offices shall play a role in developing skills in patent drafting to assist local innovators.

5. Mainstreaming utility models into the national priorities of development

Efforts will be made to promote the use of utility models to facilitate development of minor and incremental innovations that will contribute to the realization of the objectives set within the national priority areas of development and to promote local solutions to local problems.

6. Funding

Concrete action will be undertaken to facilitate funding of local innovators who will file utility models. Funding shall be provided to facilitate the development of the invention, IP protection, prototyping, and commercialization. The establishment of a 'Fund for the Promotion of Utility Models' shall be considered.

7. Awareness creation

Specific initiatives shall be undertaken to promote education on the use of utility models among artisans, single innovators, researchers, students, and small and medium-sized enterprises.

8. Promotional activities

To promote the use of utility models, initiatives shall be undertaken, including but not limited to, fairs, competitions, awards, and research grants.

9. Synergies between innovators and industry

Action shall be taken to facilitate interaction between utility model innovators and industry and to assist innovators to license or establish joint ventures with industry to facilitate mass production and commercialization of their innovations. Caution shall be exercised to safeguard the interests of innovators and enable them to benefit from their innovations.

10. Clusters of technology transfer and utility models
Efforts shall be made to ensure that all 'clusters of technology transfer' will facilitate access to local innovators in order for them to innovate around the technology acquired abroad and to secure protection of their improvements through utility models. This, in turn, should ensure that the technology is domesticated and absorbed and will prompt a new innovation hub that responds to local needs.

11. Government support and resource mobilization
Government shall make every effort to assist local innovators to use, exploit, and commercialize their innovations through effective use of utility models. Government shall also mobilize national and international support, including through facilitating technology transfer and involvement of local innovators in adapting such technologies to local needs.

countries do not need to incur costs related to the replacement of infrastructures and old technologies and even face resistance to change because they can establish new digitally related industries from scratch.[159]

Increased access and use of mobile broadband prompted an increase in the use of the internet even among poor populations and have given the illusion that Africa is catching up in terms of digital inclusion.[160] However, as the services are becoming more complex, the devices more sophisticated, and knowledge unlimited, the less privileged are facing challenges related to affordability and the ability to use those services and devices and are again risking being left behind.[161]

---

[159] Dan Ciuriak and Maria Ptashkina, 'Leveraging the Digital Transformation for Development: A Global South Strategy for the Data-Driven Economy' (*Centre for International Governance Innovation*, 3 April 2019) 4 <www.cigionline.org/publications/leveraging-digital-transformation-development-global-south-strategy-data-driven/> accessed 30 April 2023.

[160] Joseph M Kizza, 'Mobile Money Technology and the Fast Disappearing African Digital Divide' (2013) 5 AJSTID 373, 373–78; Robin Mansell, 'Digital Opportunities and the Missing Link for Developing Countries' (2001) 17 OREP 282, 293.

[161] Alison Gillwald, 'Beyond Access: Addressing Digital Inequality in Africa' (*Centre for International Governance Innovation*, 10 March 2017) 38–45 <www.cigionline.org/publications/beyond-access-addressing-digital-inequality-africa/> accessed 12 March 2019.

Therefore, although this opportunity exists, it cannot be taken for granted. Indeed, statistics show that access to computers and the internet is uneven globally: according to 2020 statistics, Africa has the lowest internet penetration rate in the world at 39.3 per cent, and the continent's share of the global number of internet users corresponds to just 11.3 per cent.[162] This is evidence enough to conclude that digital inequality still persists in the world and African populations may not benefit from opportunities emerging from the digital economy. The main challenges that Africa faces are related to lack of skills, infrastructure, and appropriate governance systems and policies.[163] Therefore, although digital technology has the potential to become a tool to empower the less privileged, if not properly capitalized it can instead perpetuate the digital divide.[164]

Promoting a balanced digitalization of Africa and ensuring a more equitable digital future requires government intervention. Governments in African countries need to craft favourable legal and regulatory frameworks and proactive policies so that people and enterprises can take advantage of the opportunities offered by the digital economy.[165] Governments need to view the digital economy as a priority and to that end develop a comprehensive plan for governance of the digital economy in order to enable the country to extract maximum benefits from it.[166] A promising example that is worth emulating is that of Rwanda, which resolved to prioritize technology, especially the digital economy, by enabling relevant policies, investing in telecommunications infrastructure, technological adoption, and education strategy, and this seems to be paying off.[167]

---

[162] 'Africa Internet Users, 2022 Population and Facebook Statistics' <www.internetworldstats.com/stats1.htm> accessed 12 March 2019; UNCTAD, 'Digital Economy Report 2019—Value Creation and Capture: Implications for Developing Countries' 12–14 <https://unctad.org/publication/digital-economy-report-2019> accessed 12 January 2020; Bangaly Kaba and Peter Meso, 'Benefitting from Digital Opportunity: Do Socio-Economically Advantaged and Disadvantaged Groups React in the Same Ways?' (2019) 22 JGITM 257, 257.

[163] Odilile Ayodele, 'The New Information Feudalism: Africa's Relationship with the Global Information Society' (2020) 27 SAJIA 67, 81.

[164] Carlos Braga, 'Development Goes Digital' (2000) 1 GJIA 23, 23–27; Bruce Mutsvairo and Massimo Ragnedda, 'Comprehending the Digital Disparities in Africa' in Bruce Mutsvairo and Massimo Ragnedda (eds), *Mapping Digital Divide in Africa* (Amsterdam University Press 2019) 13–23.

[165] Braga (n 164) 164; Elizabeth Pollitzer, 'Creating a Better Future: Four Scenarios for How Digital Technologies Could Change the World' (2018) 72 JIA SIPA 75, 75–90; Fu, Zanello, and Owusu (n 158) 34.

[166] OECD, 'Digital Economy Outlook 2017' <www.oecd.org/digital/oecd-digital-economy-outlook-2017-9789264276284-en.htm> accessed 19 July 2020; Ciuriak and Ptashkina (n 159) 5; Marta Götz, 'Attracting Foreign Direct Investment in the Era of Digitally Reshaped International Production. The Primer on the Role of the Investment Policy and Clusters—The Case of Poland' (2020) 26 J East-WestBus 131, 143.

[167] 'Rwanda Vision 2020' <www.nirda.gov.rw/uploads/tx_dce/05_-_Vision_2020__06.pdf> accessed 16 November 2018; 'The SMART Rwanda Master Plan 2015–2020' <www.minict.gov.rw/fileadmin/user_upload/minict_user_upload/Documents/Policies/SMART_RWANDA_MASTERPLAN.pdf> accessed 16 November 2018.

In view of this, there is a case for a comprehensive national strategy on the digital economy which should prioritize more investments in the requisite infrastructure, technological acquisition, reliable internet, quality standards, data security, software, hardware, and related services as building blocks for the future sustainable digital economy.[168]

Secondly, all efforts to attract foreign direct investment should prioritize measures that can attract investments in the digital economy and encourage investors not only to bring skills and technology for the benefit of their own companies but also to assist the digital transformation of the host country.[169]

Thirdly, Africa needs to take advantage of its young population and the enthusiasm of universities and research institutions to build digital skills that will facilitate participation of the continent in the new digital economy.[170] The importance of improving technological learning and building local capabilities in order to develop the absorptive capacities of developing countries and enable them to benefit from the flows of technologies from the developed world has been advocated throughout this book.[171] This assertion is even more valid now in the context of the digital economy, hence the priority of governments should be to improve digital skills in their respective countries through targeted interventions. Governments should also stimulate industry-academia linkages so that the skills imparted, and research conducted, respond to industry and market needs.

Fourthly, there is a need for targeted intervention to identify areas where the country can have a competitive advantage. In those areas, governments also need to promote research and development focused on finding solutions to specific local challenges.[172] This may be through mainstreaming digital tools to improve productivity, access, and efficiencies. Some of the areas that may require special attention in Africa due to their possible overwhelming impact and benefits include: financial services, education, health, retail, agriculture, and e-government.[173] It is also

---

[168] Pollitzer (n 165) 76–77; OECD (n 166) 128.

[169] Götz (n 166) 150–55.

[170] Mohammad Amir Anwar and Mark Graham, 'Digital Labour at Economic Margins: African Workers and the Global Information Economy' (2020) 47 ROAPE 95; Walter Matli and Mpho Ngoepe, 'Capitalizing on Digital Literacy Skills for Capacity Development of People Who Are Not in Education, Employment or Training in South Africa' (2020) 12 AJSTID 129, 129–39; Pat Cataldo, 'Building an Industrial/Academic Alliance: The Two Directions of Technology Transfer' (1989) 29 ET 33, 33–34.

[171] Ann Njoki Kingiri and Xiaolan Fu, 'Understanding the Diffusion and Adoption of Digital Finance Innovation in Emerging Economies: M-Pesa Money Mobile Transfer Service in Kenya' (2020) 10 InnovDev 67, 81.

[172] Charles F Rice, Erol K Yayboke, and Daniel Runde, 'Kenya Case Study' in Charles F Rice, Erol K Yayboke, and Daniel Runde (eds), *Innovation-Led Economic Growth: Transforming Tomorrow's Developing Economies through Technology and Innovation* (Center for Strategic & International Studies 2017) 26.

[173] Doreen Akiyo Yomoah, 'Lions Go Digital: The Internet's Transformative Potential in Africa' (McKinsey and Company 2013) 11 <https://africanarguments.org/2013/12/lions-go-digital-the-intern

suggested that governments should provide clear guidance to move away from importation of final products and hardware and prioritize instead the acquisition of software technologies that will build the basis for production and in that way start reducing dependency.[174]

Fifthly, governments should support companies or individuals that are bringing about innovations, although caution must also be exercised to avoid new monopolies that may stifle innovation. Therefore, regulations that support innovation must also be balanced enough to keep encouraging competition and more innovation.[175]

Sixthly, although technology acquisition is crucial, for it to have an impact, adequate adaptation to the context and responding to local needs will be decisive. For example, Kenya is today labelled the 'global birthplace for digital payments' because of the Mpesa system of payments.[176] However, it is worth noting that this achievement was a result of the appropriate combination of technology developed abroad, transferred to Kenya, and adapted to the local context. Indeed, the initial technological idea was developed in the United Kingdom by Vodafone, tested from 2005 in Kenya, and finally launched in 2007 after customization in order to respond to the needs of local poor consumers in Kenya.[177] The national digital plan will set the foundations to facilitate the adaptation of the technology imported from abroad in order to meet local requirements and address local needs.

The digital economy is certainly the future, and no country can avoid it. The rapid penetration of digital technologies in Africa and the fact that their implementation does not require the infrastructure that characterized the Industrial Revolution may constitute a unique opportunity for Africa to leapfrog and promote its own development. Therefore, governments need to pay particular attention to the potential of digital tools, mainstreaming them into all developmental policies in order to bridge the digital divide, reduce digital inequality, and build a digital future for Africa that will foster innovation. Since the building blocks of the digital economy reside in the technology and skills that the continent does not yet possess, it is important that all efforts to attract foreign direct investments and technology prioritize flows that will assist Africa to acquire those skills and appropriate technology.

---

ets-transformative-potential-in-africa-we-profile-the-newest-africa-report-from-mckinsey-by-doreen-akiyo-yomoah/> accessed 12 April 2019.

[174] Rasmus Lema and others, 'Renewable Electrification and Local Capability Formation: Linkages and Interactive Learning' (2018) 117 EnergyPol 326, 329.
[175] Kingiri and Fu (n 171) 81.
[176] Rice, Yayboke, and Runde (n 172) 22.
[177] ibid 72; Lema and others (n 174) 326; Bitange Ndemo and Tim Weiss, 'Making Sense of Africa's Emerging Digital Transformation and Its Many Futures' (2017) 3 AJM 328, 334–40.

h) Focus on leveraging the Fourth Industrial Revolution
As far back as 2007, the European Patent Office (EPO) enquired how IP regimes could evolve by the year 2025. The answer was that it could evolve in several directions, but four scenarios were prominent and it is worth considering them:[178]

 i) Market Rules: a scenario in which business would be the dominant driver. In such a scenario, companies dominate the patent agenda and would build powerful patent portfolios and enforce their rights, leading to an increasingly litigious world; however, an overwhelming success of the system would cause its own collapse.
 ii) Whose Game: in this scenario, it is the world of geopolitics that is the dominant driver. As a consequence, nations and cultures compete and the new entrants prevail and manage to shape the evolution of the system while the developed world fails to use IP to maintain technological superiority. Regrettably, in such a scenario, marginalization of the developing countries remains. Some progress is made however on developmental issues and technology transfer.
 iii) Trees of Knowledge: in this scenario, society is the dominant driver and, as such, criticism of the IP system grows, resulting in its gradual erosion. The main concern in this scenario is how to ensure that knowledge remains a common good, while concurrently acknowledging the legitimacy of reward for innovation.
 iv) Blue Skies: a scenario where technology is the dominant driver. To adapt to fast-paced technological progress, the patent system abandons the 'one size fits all' model and is forced to split: the former patent regime applies to classic technologies while the protection of new ones is achieved through other forms of IP protection, such as the licensing of rights. Fundamental here is the fact that technology dominates, and new forms of knowledge search and classification emerge.

In designing these scenarios EPO had already predicted that the major transformations that would dominate the first half of this century were genetics (biotechnology), nanotechnology, and robotics (artificial intelligence). More importantly, the convergence of these technologies was considered the main feature which would result in the combination of nano-, bio-, and cognitive technologies, underpinned by information technology.[179] EPO states that 'the synergistic effect of the three systems will lead to an explosion of new knowledge and new capabilities', resulting in 'the computer/communications revolution and the nano/ biology/

---

[178] Shirin Elahi, *Scenarios for the Future: How Might IP Regimes Evolve by 2025? What Global Legitimacy Might Such Regimes Have?* (European Patent Office 2007).
[179] ibid 28.

information revolution'.[180] The 'Blue skies scenario' seems to have materialized, as the world is currently dominated by technology and the tripartite combination of the three dimensions of technologies has led to the Fourth Industrial Revolution or '4IR'.

The 4IR is driven by, and stands on the shoulders of, the digital technology.[181] The 4IR is pervasive and carries along bespoke technologies such as artificial intelligence, robotics, the Internet of Things, autonomous vehicles, 3D printing, nanotechnology, biotechnology, materials science, energy storage, and quantum computing.[182]

Therefore, the 4IR is not about technology per se, but it encompasses a convergence of technologies in the physical, digital, and biological domains.[183] The previous revolutions appear to have bypassed the LDCs and these countries cannot afford to miss out on the benefits that are now looming with the new revolution. With regard to the 4IR, some predict four scenarios: 'digital inequality', 'digital divide', 'digital accretion', and 'digital harmony'.[184] The scenarios that point to the digital divide and digital inequality are pessimistic and predict the maintenance of the status quo with regard to access or use of information and communication technologies, or even worsening of the situation. Those who are pessimistic about the chances of LDCs taking advantage of the 4IR maintain that jobs will once again be diverted to the developed countries and even that the machines and robots will take the lion's share, leaving human beings jobless.[185]

Digital accretion and digital harmony would instead be desirable because they would allow adoption and introduction of new digital technologies, including mobilization of resources to promote local manufacturing; in the best scenario, digital technologies would be fully aligned with the needs and resources of society, including environmental preservation. In such positive scenarios, international trade agreements and regulations would facilitate technology transfer and adoption.

The facts on the ground seem to suggest that indeed the 4IR constitutes a unique opportunity for the LDCs to catch up and overcome the challenges they are currently facing, including poverty. Notwithstanding several obstacles, developing

---

[180] ibid.
[181] The main feature of the First Industrial Revolution was the mechanization of production hinging on the use of water and steam power; the Second Industrial Revolution brought about mass production by using electric power; the Third is the Digital Revolution that automated production thanks to electronics and information technology (semiconductors, mainframe computing, personal computing, and the internet).
[182] Klaus Schwab, 'The Fourth Industrial Revolution: What It Means and How to Respond' (*World Economic Forum*, 14 January 2016) <www.weforum.org/agenda/2016/01/the-fourth-industrial-revolution-what-it-means-and-how-to-respond/> accessed 8 October 2022; Klaus Schwab, Nicholas Davis, and Satya Nadella, *Shaping the Future of the Fourth Industrial Revolution* (Illustrated edn, Currency 2018).
[183] See the homonymous work of Klaus Schwab, who coined the term, Schwab (n 182).
[184] Pollitzer (n 165).
[185] Ratnakar Adhikari, 'The Fourth Industrial Revolution' in Sarah Aneel, Uzma Haroon, and Imrana Niazi (eds), *Corridors of Knowledge for Peace and Development* (SDPI & Sang-e-Meel Publications 2020).

and LDCs, especially in Africa, are showing innovative ways of leveraging 4IR value chains. Indeed, even where these countries may fail to actively engage activities requiring complex skills, collaboration in input, quality control, and processing of data can still be undertaken at least at the entry level, enabling some degree of their participation in the movement.[186] It is, however, expected that as time passes and skills improve, disadvantaged countries will catch up and take full advantage of the 4IR technologies.

One of the areas where 4IR technologies are already having a positive impact is in agriculture. The use of artificial intelligence and drones is already a common feature in African agriculture, and this has been successfully tried and tested in several countries. An interesting example emerges from the alternative solutions provided for by 4IR technologies that assist in overcoming the challenges in the agricultural sector in Southern Africa, where cyclical droughts (including two recent El Niño-induced droughts, in 2015–16 and 2018–19) and inadequate crops result in low productivity. This scenario is on the brink of change with the use of 4IR technologies.[187]

One mechanism to overcome the challenges is used in the biotechnology and genomics space, where a gene-editing technology known as CRISPR (Clustered Regularly Interspaced Short Palindromic Repeats) provides faster and cheaper development of drought-resistant cultivars and breeds. Also of help is the practice of precision agriculture, in which the exact quantities of water and nutrients needed by crops are provided in indoor farms with climate-controlled environments. This technology may assist farmers to break away from the limitations posed by seasons and climate that have been the main challenge for traditional farming. Sophisticated precision farming technology is now possible in the form of data and sensor technologies. In this regard, technologies such as remote sensing with satellites, unmanned aerial vehicles (UAVs or drones), or land vehicles use optical sensors to measure soil properties, including organic matter and moisture content, while the Normalized Difference Vegetation Index (NDVI) enables measurement of how plants absorb visible light and reflect infrared light: all these 4IR technologies are now available to farmers even in LDCs.[188]

4IR technologies may also assist Africa in reviving manufacturing activities. In the past, large geographic distances, lack of economies of scale in local

---

[186] Tom Simonite, 'The Pandemic Brings Some African Tech Workers Luxe Lodging' (*WIRED*, 19 May 2020) <www.wired.com/story/pandemic-brings-african-tech-workers-luxe-lodging/> accessed 12 December 2020; Dave Lee, 'Why Big Tech Pays Poor Kenyans to Teach Self-Driving Cars' *BBC News* (3 November 2018) <www.bbc.com/news/technology-46055595> accessed 9 October 2022.

[187] Mari-Lise Du Preez, '4IR and Water-Smart Agriculture in Southern Africa: A Watch List of Key Technological Advances' (South African Institute of International Affairs 2020) <www.jstor.org/stable/resrep29534> accessed 5 September 2020; Michelle Chivunga and Alistair Tempest, 'Digital Disruption in Africa: Mapping Innovations for the AfCFTA in Post-COVID Times' (South African Institute of International Affairs 2021) <www.jstor.org/stable/resrep28288> accessed 2 February 2021.

[188] Du Preez (n 187); Chivunga and Tempest (n 187).

production, the high cost of energy, and weak public-sector capacity made it difficult for African countries to engage in manufacturing.[189] The 4IR is democratizing production by expanding the number of producers, including small and medium-sized enterprises, and dematerializing it through digital production, making manufacturing much easier, cheaper, and more sustainable. This means that there is a unique opportunity for Africa to foster manufacturing activities by leveraging the opportunities provided by the 4IR technologies. For example, the production of sophisticated processing machinery to be used in the food and beverages industry will increase in order to satisfy the demand of the middle class in exponential expansion. Artificial intelligence may help in tracking products along the supply chain, improve the efficiency of food processing and packaging, and reduce food waste. Lastly, renewable energy technologies such as solar panels and batteries may improve the competitiveness of African manufacturing by reducing the cost of electricity.[190]

The 4IR technologies will also be a powerful tool to boost intra-African trade. Leveraging the emerging technology also has the potential to facilitate economic integration and a more inclusive framework of trade and to enable economies of scale through the use of ecommerce channels. Trade in Africa is also poised to grow because digitalization has the potential to transform its supporting processes, rendering them more efficient and assisting in the reduction of the prices of goods and services, which will have a positive impact on poor consumers on the continent.

Notwithstanding this optimism, which is reiterated even at this juncture, it is worth acknowledging that there are still some hurdles, such as the issues of lack of skills, absorptive capacities, unemployment caused by automation, and threats of cybercrime which may be felt with relatively higher intensity in Africa due to its vulnerability to these shocks.[191] The LDCs face threefold challenges to take advantage of the 4IR, according to Adhikari:[192]

i) Accessibility: a limited number of countries have a tight hold over technologies required by LDCs, ring-fencing them though patent protection, and enterprises and individuals in LDCs fail to use 4IR technologies due to absence or deficiency of infrastructure—some of the infrastructural issues are related to power supply, which is crucial for access to internet.

ii) Affordability: use of 4IR technologies by consumers may be hindered by excessive prices charged for the use of these technologies or devices and

---

[189] Wim Naudé, 'Brilliant Technologies and Brave Entrepreneurs: A New Narrative for African Manufacturing' (2019) 72 JIA SIPA 147–48.
[190] ibid 149.
[191] ibid.
[192] Adhikari (n 185) 52.

this needs to be looked at; affordability of internet, devices, and technologies is the condition sine qua non to enable consumers to enjoy existing technologies

iii) Application: the chances of effective utilization of the technologies for the benefit of consumers, be they firms or individuals, require appropriate policies, regulations, and skills.

The government action for skills development and prioritization of STEM has been illustrated throughout Section B.2 on 'Government Interventions to Develop Absorptive Capacity'. As far as the policies are concerned, government decisions to provide access to the internet even to the least favoured individuals; the granting of tax rebates to industries that make use of 4IR technologies; deregulating ICT services or granting tariff reductions on the importation of ICT materials; and liberalizing foreign ownership of companies dealing with 4IR-related technologies may stimulate improved adoption and application of 4IR technologies in developing and LDCs.[193]

What is however required in this case is for the governments to adopt consistent and adequate policies to address those threats. But this will not happen easily, and the path is not straightforward. The Digital Transformation Strategy for Africa, 2020–2030, adopted by the African Union, appears to lead in the right direction in efforts to usher Africa into the Digital Revolution and 4IR. For this Strategy:

> Africa presents a sea of economic opportunities in virtually every sector, and the continent's youthful population structure is an enormous opportunity in this digital era and hence the need for Africa to make digitally enabled socio-economic development a high priority.

The Strategy finds that the current dynamics offer a leapfrogging opportunity for the continent of Africa, hence the need for a coordinated response to reap the benefits of the 4IR. To generate the desired impact, these efforts should achieve greater depth at the national level so that each government may play an active role in creating an enabling environment for the 4IR to establish roots and avoid the continent being once again bypassed by the new industrial revolution that the world is witnessing. Some of the measures that African governments should undertake include:

i) Prioritization of human skills development in order to sustain the digital economy and the needs of the 4IR;

---

[193] ibid 51.

ii) Promoting the development of managerial and entrepreneurial skills in Africa's emerging technology hotspots and entrepreneurial ecosystems;
iii) Investing in digital infrastructure and efforts to reduce the digital gap;
iv) Supporting growth of venture capital finance to fund new high-tech ventures, particularly in priority areas for Africa such as manufacturing, agriculture, renewable energies, transportation, and communications;
v) Improving regulatory environment to facilitate the growth of digital business and the emergence and exploitation of 4IR technologies;
vi) Developing targeted measures to promote the flow of technologies to the continent that may catapult Africa into the 4IR;
vii) Developing the appropriate legal framework to address the issue of IP in the digital economy.

The 4IR will not flourish in the continent by chance but requires deliberate action from governments, the private sector, and academia. The primary element in this equation is the consciousness on the part of the African stakeholders that there is an opportunity for Africa to catch up in economic and technological progress through appropriate use of 4IR technologies. Secondly, there needs to be decisive action to put in place adequate policies and measures to enable the continent to participate actively in the 4IR.

C. Conclusion

In pursuing ways of promoting technology transfer to LDCs, especially through the implementation of Articles 7 and 66.2 of the TRIPS Agreement, there has been a dominant trend of viewing the responsibilities of facilitating transfer of technology to LDCs as a one-way set of obligations that falls only on developed states. What about the LDCs themselves? What steps have they taken to attract and retain technology with a view to developing their own technology base? Drawing from the lessons learnt from the successful experiences of East Asian states in promoting technological progress and innovation, it was made clear that there is a need to create an enabling environment to trigger inflows of technologies in the recipient states through the adoption of appropriate interventions. First and foremost, the important role of defining their development priorities, and the technological areas that can sustain these, falls on the shoulders of African LDCs. Thereafter, LDCs are called upon to develop technological competencies and absorptive capacities in their respective states. This can be achieved through appropriate human development capital; dedicated institutional frameworks; adequate funding allocated to research, development, and education; indigenization of foreign technologies through utility models; and the socialization of science and technology. This will enable them to promote technological learning that will lead to the development of capacities for reverse engineering and adaptation of technologies to the local context and ultimately innovation that will be based on local needs.

A special focus is placed on the use of utility models to facilitate technological learning, improvement of absorptive capacities, and adaptation of imported technologies to local needs. The huge benefits that could accrue from the use of utility models contrast with the scarce use of this IP category on the African continent. This requires that national strategies and policies be adopted to fast-track their uptake by governments and, most importantly, by local innovators to facilitate the transformation of the minor and incremental innovations that arise throughout the informal economy into valuable intangible assets with adequate legal protection.

Also of particular relevance is the exploitation of the opportunities offered by the 4IR: If Africa was bypassed by the first three revolutions, the current revolution appears to be a unique opportunity not to be missed by the continent. Therefore, appropriate policies ought to be adopted both at the continental and regional levels, and also at the national level, to enable African countries to leverage the population dividend that currently favours the continent, especially due to predominance of youth, in order to catch up on technological progress. For Africa to fully benefit from the digital economy and the 4IR, human skills development, investment in infrastructure, and funding are crucial and deserve special attention. The focus of transfer of technology should therefore be to facilitate access to technologies that will enable Africa to fully participate in 4IR and to reap the benefits deriving from it.

8
Final Conclusions and Recommendations

A. Introduction

This book has postulated from the outset that Africa, being the cradle of humanity and ingenuity, ought to reclaim its space in innovation. Through expository methods evidence is unveiled to demonstrate that Africa undertook the preliminary and fundamental steps that decisively contributed to shaping the world technological progress witnessed today. However, it appears as though the 'out of Africa' process not only relocated men to other continents, but also removed the innovative capabilities to which Africa was home.

One of the paradoxes of our time is of why African states are trailing behind other states if Africa was so advanced previously. This book has undertaken its own 'out of Africa' process to understand the reasons that brought African technological progress to a standstill. Some endogenous factors seem to have played a major role in freezing the progress of Africa in the technological domain. These include strong cultural behaviour that defines 'African conservatism' and what appear to be brakes and blockages to technological progress.[1] These factors have also shaped the African approach to the intellectual property (IP) system and the reluctance to use it as a tool to protect and incentivize knowledge. More disconcerting is how this behaviour has distanced the continent from others and worked against its own people.[2]

In trying to understand the barriers that Africa is facing in its attempts to access and use technology and foster innovation, this book raises three issues that are elaborated on throughout. The first is an attempt to understand the root causes and the current impediments to technological progress. The second interrogates the current global IP system and enquires how it could be reformed to allow African states to access and develop their own technology and promote innovation. Thirdly, the book seeks to understand the role that African states can play in improving

---

[1] Ralph A Austen and Daniel Headrick, 'The Role of Technology in the African Past' (1983) 26 ASR 163, 171–74; George Ovitt, 'The Cultural Context of Western Technology: Early Christian Attitudes toward Manual Labor' (1986) 27 TechnoCult 477, 480.

[2] Pasquale Joseph Federico, 'Origin and Early History of Patents' (1929) 11 Journal of the Patent Office Society 292, 293–94; Christopher May and Susan K Sell, *Intellectual Property Rights: A Critical History* (Lynne Rienner Publishers 2006) 72; Christopher May, 'The Hypocrisy of Forgetfulness: The Contemporary Significance of Early Innovations in Intellectual Property' (2007) 14 RIPE 1, 2; Pamela O Long, 'Invention, Authorship, "Intellectual Property," and the Origin of Patents: Notes toward a Conceptual History' (1991) 32 TechCult 846, 848.

access to technology and developing absorptive capabilities that will enable them to assimilate and adapt foreign technologies to address local needs and promote home-grown innovation. Apart from uncovering history to find the reasons behind Africa's technological stagnation through the first question, the subsequent queries enable this book to formulate some ideas on the way forward. Indeed, with regard to the second issue, two objectives are articulated, these being, firstly, to determine the philosophical and legal obstacles to Africa's participation in the global innovation process and access to technology; and, secondly, to address these obstacles with a view to recommending new ways in which Africa can be enabled to take part in the global innovation process and obtain access to technology. With regard to the third issue, the objective is to explore how Africa can also take action to attract foreign technologies and create an enabling environment for their absorption and indigenization in order to address local needs and foster local innovation.

These objectives were fully achieved, first by identifying the main philosophical justifications to the IP system and targeting the dominant one, the natural rights or labour theory, that was inspired by the works of the philosopher John Locke. Shortcomings in the interpretation of this theory are exposed mainly due to the fact that it had been wrongly applied to justify the establishment of strong IP regimes. It is consequently established that it is this erroneous approach to the theory that is the culprit in barring Africa from access to technology. The book then proposes a reinterpretation of the theory in light of the principles of justice and cosmopolitanism in conjunction with the rights-based approach to development. With regard to the legal obstacles, the book notes the absence of a global regime to promote technology transfer and foster innovation and observes piecemeal attempts to address the issue that are a clear manifestation of fragmentation of the regime. Accordingly, it proposes streamlining the system through the adoption of a single legal international instrument. It is also emphasized that the existence of a UN institution that aims at focusing on technology transfer and innovation— the Technology Bank for Least Developed Countries (LDCs)—is to be leveraged. Finally, it is proposed that making use of the existing Dispute Settlement Mechanism (DSM) under the WTO is the most efficient way of dealing with disputes that may arise in this domain.

Notwithstanding the findings outlined, it became evident that fostering innovation requires introspection on the part of African states and concerted action to be taken to achieve the objective of fostering innovation. One of the fundamental issues to be addressed is creating absorptive capacities in order to take advantage of foreign technologies and to promote home-grown African innovation.

This chapter provides a summary of the main findings of this book and is structured as follows: it discusses firstly how to overcome the philosophical obstacles to Africa's participation in the global innovation process and access to technology and to that end it recommends a reinterpretation of the Locke's theory on the duty of charity with a view to transforming it into a binding legal obligation so that it is effective in assisting developing states to substantiate their claim for integration into the global innovation system. Having identified the Agreement on Trade-related Aspects of Intellectual Property Rights (TRIPS Agreement), especially Article 66.2, as the main legal instrument that can legally ensure transfer of technology to Africa, a thorough analysis of how its use can be maximized is conducted and proposals for concrete actions to be undertaken are provided. This is followed by a summary of the main submission of this book, which is the proposal for establishment of an Agreement on Trade-Related Issues on Technology Transfer and Innovation (TRITTI). To conclude, a thorough analysis of the factors that can assist in creating an enabling environment to facilitate flows of technology to recipient states is provided. Drawing from that discussion, recommendations on how to address the issues raised are presented.

B. Summary of the Findings and Recommendations

1. Overcoming the Philosophical Obstacles to Africa's Participation in the Global Innovation Process and Access to Technology

There is a legitimate claim from labourers to own the fruits of their labour, which is philosophically sustained by the natural rights or labour theory that was developed by John Locke.[3] The claim to one's fruits is based on the fact that one has laboured and added value to the products.[4] This claim, originally conceived in the context of tangible property, was later translated to apply to intangible property as well.[5] This is because it was found that intangible assets are much more vulnerable than tangible assets as they are easier to access, copy, and use without consent. Therefore, IP rights appear to be the only mechanism that intellectual labourers can use to secure control of the proceeds of their work.[6] However, this claim was pushed too far and ultimately became the justification for strong IP rights that barred access to the achievements of human ingenuity by developing states, including Africa.[7] For these reasons, this book found that the interpretation of Locke's theory of natural rights is too restrictive as it only focused on proprietary rights and the strengthening of the IP regimes as the appropriate methods to achieve protection

---

[3] John Locke, *Two Treatises of Government* (Peacock 1689); Robert Nozick, *Anarchy, State, and Utopia* (Reprint edn, Basic Books 2013) 178–82.
[4] David B Resnik, 'A Pluralistic Account of Intellectual Property' (2003) 46 JBE 319, 322.
[5] Wendy J Gordon, 'A Property Right in Self-Expression: Equality and Individualism in the Natural Law of Intellectual Property' (1993) 102 YaleLJ 1533, 1549; Lawrence Becker, 'Deserving to Own Intellectual Property' (1993) 68 Chic-KentLR 609, 610.
[6] Bryan Cwik, 'Labor as the Basis for Intellectual Property Rights' (2014) 17 ETMP 681, 689–90.
[7] Gordon (n 5) 1540; Matthew E Fischer, 'Is the WTO Appellate Body a "Constitutional Court"? The Interaction of the WTO Dispute Settlement System with Regional and National Actors' (2009) 40 GJIL 291, 4.

of one's intellectual labours. However, it is submitted that this can be corrected by reframing Locke's theory and blending it with other equally noble principles, such as the principles of justice, cosmopolitanism, and a rights-based approach to development.

The basis for reframing the natural rights theory into a less restrictive mechanism is a consideration of two moral duties that fall on the shoulders of the rights-owners, also enunciated by Locke in the context of the natural rights theory: the duty of charity, which imposes the obligation to provide for those who are in need, and the 'prohibition of waste', which forbids those who possess rights or goods to waste what they have created or produced.[8] Locke had indeed set conditions for the enjoyment of proprietary rights in Chapter 5 of the *Two Treatises of Government* through the 'enough and as good' and the spoilage (waste) provisos. The 'enough and as good' proviso allows creators to have rights to their own endeavours but only if such grant of property does not deprive others of the opportunity to also create or to benefit from existing knowledge.[9] The waste proviso is even more relevant because it prohibits waste of what human ingenuity has produced to satisfy a demand due to proprietary claims.[10] It appears that Locke's theory has a component of benevolence and inclusion that enables limitation of property claims in some circumstances with the view to promoting justice in the world. Apparently, this component was ignored in the case of the IP system. Therefore, it is suggested that more research should be undertaken to understand how Locke would have applied property regimes in the context of intangible assets where the imperative is to ensure that all of humanity derives benefits.[11] This book contributes to this debate.

The starting point is to identify technology as the main tool for development that is lacking in the developing world.[12] However, there is 'enough and as good' of technology that could be channelled to the developing world to meet its needs. For instance, it has been argued that the technology that is necessary to meet the basic needs of the developing world, such as electricity production, water and sanitation, agriculture, communication, and several areas of basic health, is widely available in developed states, but that the gap in terms of access and use of that technology is

---

[8] Gordon (n 5) 79–80; Robert P Merges, *Justifying Intellectual Property* (Harvard University Press 2011) 1570.
[9] Gordon (n 5) 1570.
[10] Gordon Hull, 'Clearing the Rubbish: Locke, the Waste Proviso, and the Moral Justification of Intellectual Property' (2009) 23 PAQ 67, 81.
[11] ibid.
[12] Anja Breitwieser and Neil Foster-McGregor, 'Intellectual Property Rights, Innovation and Technology Transfer: A Survey' (The Vienna Institute for International Economic Studies 2012) 88, 47 <https://wiiw.ac.at/p-2646.html> accessed 24 April 2023; Hans Duller, 'Role of Technology in the Emergence of Newly Industrializing Countries' (1992) 9 ASEAN EcoBul 45, 45–54; Saon Ray, 'Technology Transfer and Technology Policy in a Developing Country' (2012) 46 JDA 371, 371; Keith Maskus, 'Encouraging International Technology Transfer' [2004] IPRSD 1, 7; M Scott Taylor, 'TRIPS, Trade, and Technology Transfer' (1993) 26 CJE/Revue canadienne d'économie 625, 625–37; Engwa Azeh Godwill, 'Science and Technology in Africa: The Key Elements and Measures for Sustainable Development' (2014) 14 GJSFR: G BIO-TECH & GEN 16.

widening between developed states and LDCs.[13] The widening of this gap is mainly attributed to proprietary approaches that are intrinsic to the IP system that has prevented developing states from benefiting from the technological progress attained by humanity.[14]

The book therefore demonstrates that, philosophically, it is possible to limit proprietary claims in order to attend to equally noble causes such as justice and fairness.[15] Therefore, by elaborating on the Lockean provisos, namely the duty of charity and the prohibition of waste, an ethical responsibility on developed states to transfer technology to developing states and promote innovation is sustained. These moral obligations must be taken into account by policymakers during the decision-making and norm-setting processes. This will allow them to translate those ethical responsibilities into a mechanism that will contribute to reshaping the IP system in order to avoid exclusive focus on the protection of IP rights, but instead to turn it into a tool that can promote justice and welfare for everyone in the world. Accordingly, the theory of cosmopolitanism is applied, resulting in the recognition of every human being as a 'unit of moral concern' entitled to see his claims addressed at the global level.[16] Indeed, the cosmopolitan theory has developed the 'equal opportunity principle' based on the concept that everyone's claims be treated equally, regardless of their nationality.[17] Therefore, cosmopolitanism justifies international collaboration to meet people's basic needs and improve their standard of living, regardless of their nationality. Through this theory, it is philosophically justified that developed states and their citizens render support to poor states to promote innovation for the benefit of their people. What this book emphasizes is that support must be geared towards promoting effective development through the promotion of creativity and innovation. This can be achieved by assisting developing states and LDCs to develop absorptive capabilities in order to catch up in their technological progress and innovation. The development of technological capabilities in developing states is addressed in detail in Chapter 7 and recommendations provided, focusing on the establishment of government institutional frameworks devoted to maximizing the use of science, technology, and innovation to promote innovative capabilities; increasing funding for research and

---

[13] UNCTAD, *Transfer of Technology and Knowledge-Sharing for Development: Science, Technology and Innovation Issues for Developing Countries* (United Nations 2014) 3; NJ Udombana, 'The Third World and the Right to Development: Agenda for the Next Millennium' (2000) 22 HRQ 753, 783.

[14] Udombana (n 13) 783; Surendra J Patel, 'The Technological Dependence of Developing Countries' (1974) 12 J ModAfrStud 1, 1–5.

[15] Hull (n 10) 87; Gordon (n 5) 1551.

[16] John Sellars, 'Stoic Cosmopolitanism and Zeno's "Republic"' (2007) 28 HPT 1, 6–7; Thomas Pogge, *World Poverty and Human Rights*, vol 19 (Cambridge University Press 2002) 169 <www.cambridge.org/core/journals/ethics-and-international-affairs/article/abs/world-poverty-and-human-rights/A647319E9BEE481BAABCADD0B982D89D> accessed 1 May 2023; Carl Knight, 'In Defence of Cosmopolitanism' (2011) 58 Theoria: JSPT 19, 19.

[17] Knight (n 16) 20–21.

development; developing human capital; and adopting public policies supportive of technology adoption, including by encouraging the use of utility models.

The book also addresses the argument that developed states are unable to force companies that own technologies to transfer them to LDCs.[18] In that regard, it is demonstrated that the cosmopolitan ideals can be used to provide the fundamental philosophical underpinnings to persuade business leaders to comply with their moral obligation to render support. This is especially true with regard to the states in which they are operating because, with the right support, those states may access technologies and create a solid technological base and capability to pursue technological progress and development.[19] Conscious of the fact that a moral obligation is not enough to compel entrepreneurs to transfer technology to developing states, the need to translate these moral obligations into legal instruments (legal cosmopolitanism) to render them enforceable is advanced here.[20] It is for this reason that the establishment of a legally all-encompassing international instrument is proposed: the Agreement on Trade-Related Issues of Technology Transfer and Innovation (TRITTI) that would set minimum standards that governments should further domesticate at the national level to force or persuade the private sector to collaborate in the transfer of technologies to developing states. This is feasible because the transfer of technology would be given in exchange for the necessary safeguards to their IP rights through national and international legal frameworks, thus assuring them that their technologies are safe in developing states.

Another important tool to sustain the quest by developing states to access technology and promote innovation is the right to development.[21] The right to development requires collective action by states to address collective concerns.[22] Therefore, it has acquired legal status as a human right internationally through the Declaration on the Right to Development adopted by the UN General Assembly in 1986.[23] The right to development was also recognized by the Vienna Declaration and Programme of Action adopted by the UN World Conference on Human Rights held in Vienna in 1993 as a universal and inalienable right and an integral part of fundamental human rights; hence, it calls upon the international community

---

[18] European Union, 'Report on the Implementation of Article 66.2 of the TRIPS Agreement' (WTO) Doc IP/C/R/TTI/EU/3 <https://docs.wto.org/dol2fe/Pages/FE_Search/FE_S_S009-DP.aspx?language=E&CatalogueIdList=288475,288461,288378,288403,288245,288236,288039,288037,288041,279272&CurrentCatalogueIdIndex=0&FullTextHash=&HasEnglishRecord=True&HasFrenchRecord=True&HasSpanishRecord=True> 30 October 2023.
[19] Thomas Maak and Nicola M Pless, 'Business Leaders as Citizens of the World. Advancing Humanism on a Global Scale' (2009) 88 JBE 537, 539–45.
[20] Seyla Benhabib, 'The Legitimacy of Human Rights' (2008) 137 Daedalus 94, 97; Pogge (n 16) 49.
[21] Bård-Anders Andreassen, 'On the Normative Core of the Right to Development' (1997) 24 ForumDevStud 179, 187–89.
[22] Sakiko Fukuda-Parr, 'The Right to Development: Reframing a New Discourse for the Twenty-First Century' (2012) 79 SocRes 839, 842.
[23] United Nations Declaration on the Right to Development (adopted 4 December 1986), A/RES/41/128; Ragnar Hallgren, 'The UN and the Right to Development' (1990) 22/23 Peace Research 31, 31–41.

to cooperate in order to realize these rights.[24] By virtue of the Declaration on the Right to Development, every human person and all peoples are entitled to participate in, contribute to, and enjoy economic, social, cultural, and political development, in which all human rights and fundamental freedoms can be fully realized. The Declaration on the Right to Development therefore explicitly highlights the need for an international enabling environment that can foster development and thus calls on states to formulate appropriate international policies to address inequalities.[25] Given that the relevance of the right to development is set out, the task is now to determine how the right to development can best be used to support the calls for a more balanced IP system that fosters innovation in developing states. One practical way of doing this is to align African IP laws and policies with the development priorities of their respective states. It is suggested that African states should expressly invoke in their IP laws the issue of human development as one of their objectives. This will allow their legislation to be development-orientated, enable national courts to take into account the objectives of development in settling IP disputes, and promote a human development approach in the implementation of international IP obligations.[26]

The right to development finds its full realization through the rights-based approach because it reframes development into an entitlement.[27] The rights-based approach sustains the claim for the satisfaction of basic needs in the form of a right rather than charity, and to that end it grants legal and ethical authority to developing states.[28] Therefore, the claims for better access to technology and promotion of innovation that were discussed can also find solid foundations in the rights-based approach to development. These philosophical underpinnings were subsequently translated into more solid legal foundations to sustain the quest by developing states for better access to technology and innovative capabilities as highlighted in Section B.2 on 'Maximizing the Use of Article 66.2 to Promote Technology Transfer to Africa'.

2. Maximizing the Use of Article 66.2 to Promote Technology Transfer to Africa

Transforming the right to development into an entitlement as conceived by the rights-based approach requires a strong substantive international legal

---

[24] Vienna Declaration and Programme of Action (adopted 25 June 1993) Doc A/CONF.157/23.
[25] Fukuda-Parr (n 22) 840.
[26] J OseiTutu, 'Prioritising Human Development in African Intellectual Property Law' (2016) 26 WIPO J 32–33.
[27] Arjun Sengupta, 'On the Theory and Practice of the Right to Development' (2002) 24 HRQ 837, 846; Paul Gready, 'Rights-Based Approaches to Development: What Is the Value-Added?' (2008) 18 DevPrac 735, 737.
[28] Brigitte I Hamm, 'A Human Rights Approach to Development' (2001) 23 HRQ 1005, 1026.

foundation.[29] The legal instrument with the requisite power is undoubtedly the WTO TRIPS Agreement. Hence, the obligations that fall on the shoulders of developed states as a result of the commitments enshrined in TRIPS are explored in this book. Article 7 of the TRIPS Agreement establishes a global responsibility to promote technological development and dissemination of technology. This applies for the benefit of producers that are invariably located in developed states, as well as users of the technologies that are usually found in poor states. Accordingly, Article 66.2 of TRIPS assigns to the developed member states the task of providing incentives to producers of technology located in their territories with a view to persuading them to share their technology with LDC members in order to enable those states to create a sound and viable technological base. Regrettably, there is no evidence of concrete implementation of Article 66.2 and, as a result of this, there is hardly any technology—or only an insignificant amount—being transferred to LDCs.[30] However, since some initiatives and incentives to promote technology transfer appear to have been adopted by developed states, it is recommended that those initiatives be specifically linked to the incentives that triggered them to ensure transparency, consistency, and effective transfer of technology to LDCs. To further ensure effectiveness, it is proposed that LDCs should establish 'clusters of technology transfer' on a geographic or sectoral basis. This would facilitate the identification of necessary technologies, the providers and users thereof, in larger territories, hence promoting economies of scale that would attract investors from the developed world. Additionally, the establishment of an online tool in the form of a database that could match the priorities and needs in terms of technology set by the LDCs and the incentives and support that could be provided by the developed states is proposed. Noting the lack of adequate coordination and administration of issues related to transfer of technology, the proposal advanced is for the establishment of institutional arrangements both upstream and downstream. With respect to developed states, specific focal points of technology transfer could more effectively gather, analyse, assemble, and systematize data on technologies transferred to LDCs. On the part of LDCs, national institutional frameworks could be established to monitor the technologies that are purportedly being transferred with a view to developing coherent strategies to exploit them for the benefit of their territories. A sample of 'Guidelines for the Field Impact-Assessment of the Technologies Transferred in the Context of the Implementation of Article 66.2 of TRIPS' that the focal points could use when undertaking field-impact assessment

---

[29] Jaakko Kuosmanen, 'Repackaging Human Rights: On the Justification and the Function of the Right to Development' (2015) 11 JGE 303, 304.

[30] Nefissa Chakroun, 'Using Technology Transfer Offices to Foster Technological Development: A Proposal Based on a Combination of Articles 66.2 and 67 of the TRIPS Agreement' (2017) 20 JWIP 103, 4; Jayashree Watal and Leticia Caminero, 'Least-Developed Countries, Transfer of Technology and the TRIPS Agreement' 15–22 <www.wto.org/english/res_e/reser_e/ersd201801_e.pdf> accessed 24 April 2018.

studies is provided in the book. Finally, it is acknowledged that LDCs may not, on their own, be in position to fully take advantage of Article 66.2, as is currently the case. Therefore, what is suggested is the establishment of a WTO Advisory Centre for Technology Transfer and Innovation (ACTTI) in order to assist LDCs to better identify, acquire, and exploit the flows of technology from developed states.

3. The Proposal for Establishing an International Treaty on Technology Transfer

The proposed new approach to the implementation of Article 66.2 will inevitably bring new dynamics and make a difference to the current situation. Admittedly, however, the provision still faces a number of challenges in light of the fact that rights and obligations of duty-bearers and recipient states, as well as remedies where obligations are not met, are not clearly defined. This is further exacerbated by the lack of a single, universal instrument and institution to oversee international transfer of technology as the system is currently marred by fragmentation. Consequently, the proposal is the adoption of an international treaty on transfer of technology entitled 'Agreement on Trade-Related Issues of Technology Transfer and Technology and Innovation' (TRITTI). It is proposed that the same be housed in the WTO for the following reasons: its strategic position on global trade issues; the crucial role that the WTO has played in uplifting the relevance of IP globally through the TRIPS Agreement, which has specific provisions dealing with technology transfer such as Articles 7, 8, and 66.2; and also for having established the DSM. Technology transfer transactions are intrinsically a trade issue, hence any attempt to address issues related to technology transfer will inevitably affect trade and have an impact on the trade system that is governed by the WTO. Of particular relevance is the special and differential treatment that the WTO grants to developing states to enable them to create a sound and viable technological base. Apart from the transitional periods that were set to last up until 1 July 2034 for TRIPS in general, and until 1 January 2033 for pharmaceutical-related patents in order to assist those states to catch up technologically, the preamble to TRIPS and Article 8(1) specifically mention the differential treatment for the same purpose. Trade-related measures such as incentives, subsidies, or any other initiative that may give special treatment to developing states and LDCs would best be located in the WTO as the main negotiating forum, especially as this may have an impact on technology transfer.

The likelihood of success of negotiations on a global treaty on technology transfer seems to be high and timely due to several factors. First of all, the context of TRITTI is different from the context that led to the failure of negotiations for the establishment of the UNCTAD Draft Code of Conduct on Technology Transfer in 1985. This is because the geopolitical context is no longer characterized by the same negotiating groups of the past, and even the subsisting groups, such as Group B or the Group of 77, have changed approach on several issues. For example, previous topical issues such as applicable law, the legal character of the proposed treaty, technical assistance, and conflict resolution are now viewed through different lenses. Furthermore, areas of public interest have evolved and there are now new areas, such as plant variety protection and climate change, with a significant bearing on technology transfer. Finally, the proposed TRITTI will adopt a new approach with regard to one of the most difficult issues on technology transfer: the challenge of persuading the private sector to share the technologies that they own with LDCs. Under the proposed new international legal instrument, states would be required to enact legal frameworks that establish regulatory benefits for the private sector within their national trade system in exchange for their facilitating transfer of technologies to LDCs. For example, member states, especially those in the developed world, could tie the granting of incentives to certain conditions as a mechanism to persuade companies willing to participate in government tenders related to contracts to be executed in LDCs, to transfer technology.

Although there are difficulties in negotiating a treaty under the WTO, especially considering the hurdles witnessed in the context of the Doha negotiations, the recent success in concluding the Trade Facilitation Framework is encouraging. Further, the king-makers of the WTO system, namely the European Union and the United States are now in a better position to consider issues related to transfer of technologies than ever before. This is because the content of the new multilateral regime on technology transfer may mimic the content of some of the most recent international legal instruments that were negotiated and agreed upon, especially on climate change, which have established new principles that coincide with the proposals related to TRITTI. Ultimately, the TRITTI could finally put to rest the debate regarding the creation of the right balance between the strong protection of IP rights advocated by the developed world and facilitating access to technology with a view to promoting technological progress that is demanded by the less privileged states. It would also constitute redress of the imbalances that were created by TRIPS that favoured the developed states and failed developing states in their quest to access the advantages of the global trade system, including through access to technologies.

4. Enabling Environment to Facilitate Flows of Technology in Recipient States

The misfortunes experienced by African states regarding adoption, development, and use of technologies cannot be attributed solely to external factors. Cultural approaches and misguided policies slowed the pace of adopting technology and in many cases stifled innovation on the continent. The case of the Ethiopian emperor who refused the adoption of certain technology of great benefit to his people because he found it useless for himself is just one example of the distorted approach of African rulers to the issue of technology.[31] Such an approach impeded the development of adequate mechanisms to foster innovation on the continent. Accordingly, this requires LDCs to create the appropriate enabling environment for technologies to flow into their respective states and, most importantly, to facilitate the development of technological capabilities that will accelerate technological learning and absorption of the acquired technologies.

One important aspect is the ability to mainstream IP into the processes of development in general, and more specifically, into industrial development and innovation. Therefore, it is recommended that overall, African states should consider mainstreaming IP into their national development policies. The emphasis should thus shift focus away from the view that the promotion of IP and more specifically, the protection and enforcement of IP, is an end in itself.

Noting the important role that science, technology, and mathematics can play in fostering innovation becomes a matter of concern because the innovation systems in African states are inadequate. This is because, notwithstanding the initiatives undertaken to establish the systems, they are not contextualized and are apparently simply a reproduction of patterns that are trending on the continent.[32] It is therefore suggested that ministries of science and technology in Africa be established taking into account the specific context of each state in order for them to effectively contribute to the development of their respective states. It is also noted that the cross-cutting nature of science and technology is contrasted against the vertical structure of government departments, which places the ministry of science and technology hierarchically at the same level as other ministries. It is therefore recommended that issues related to science, technology, and innovation receive special attention, including by placing the institution dealing with them under the office of the prime minister or president, in order to enable government to mainstream innovation in all strategies, policies, and plans.

The other issue discussed is the scarcity of funding for research and development in Africa. It is reported that African states hardly ever reach the threshold of 1 per cent of GDP funding on research and development in their respective territories as pledged by the heads of state and government at the African Union gatherings.[33] Two solutions to the issue of underfunding of research and development are flagged in this book: an increase in the government contribution to match the pledges made by the heads of state and government at the African Union

---

[31] Merid Wolde Aregay, 'Society and Technology in Ethiopia, 1500–1800' (1984) 17 IJES 127, 129.
[32] UNECA, 'Assessing Regional Integration in Africa VII: Innovation, Competitiveness and Regional Integration' 86 <https://repository.uneca.org/handle/10855/23013> accessed 8 October 2017.
[33] 'Only 3 African States Spend More Than 1% of GDP on Research & Development | ReSAKSS' <www.resakss.org/node/6285> accessed 6 May 2023; Decision of the Executive Council of the African Union at the Eighth Ordinary Session (adopted 21 January 2006) Doc EX.CL/Dec.254 (VIII).

to funding devoted to research and development and the benchmarks set by the United Nations Educational, Scientific and Cultural Organization (UNESCO) with regard to thresholds of funding for education; and, secondly, devising innovative ideas such as collaboration with the private sector and empowering universities to derive financial benefits from the outputs of their research.

The crucial issue of the development of human capital as the means to enable technological learning is also addressed. Several studies undertaken in different parts of the world have demonstrated the direct correlation between the level of human capital development and technological development and economic growth.[34] It is observed that the most efficient way of developing human capital is through formal education.[35] In assessing the situation on the African continent, it emerges that African states manage this issue haphazardly and that the efforts to develop human capital are jeopardized by inadequate funding. Hence, it is recommended that African states strive to meet the UNESCO thresholds for educational funding in order to facilitate the improvement of education systems and, consequently, the development of human capital on the continent, which is crucial for the development of technological capabilities.

The role that science, technology, engineering, and mathematics (STEM) education can play in setting the basis for technology and innovation to thrive is also explored. It is noted that the trend in African education systems has been that of favouring the teaching of the arts to the detriment of mathematics.[36] A positive trend is the recent and positive shift in some states where teaching of STEM subjects is being strengthened at all levels of education, and a compulsory ratio in favour of those subjects at the tertiary education level has been introduced. This is evidenced by the case of Nigeria, which has imposed a 60:40 ratio, while Ethiopia has imposed a ratio of 70:30.[37] This trend is to be emulated by other African states with a view to changing the current scenario.

---

[34] Jimmy Alani, 'Effects of Technological Progress and Productivity on Economic Growth in Uganda' (2012) 1 PEFin 14, 14–23; Muhammad Amir, Bilal Mehmood, and Muhammad Shahid, 'Impact of Human Capital on Economic Growth with Emphasis on Intermediary Role of Technology: Time Series Evidence from Pakistan' (2012) 6 AJBM 280, 280–85; Khalafalla Ahmed Mohamed Arabi and Suliman Zakaria Suliman Abdalla, 'The Impact of Human Capital on Economic Growth: Empirical Evidence from Sudan' (2013) 4 ResWorldEcon 43, 43–53; Rajabrata Banerjee and Saikat Sinha Roy, 'Human Capital, Technological Progress and Trade: What Explains India's Long Run Growth?' (2014) 30 JAE 15, 15–31; Manuel Mendes de Oliveira, Maria Costa Santos, and BF Kiker, 'The Role of Human Capital and Technological Change in Overeducation' (2000) 19 EconEduRev 199, 199–206; Jong-Wha Lee, 'Human Capital and Productivity for Korea's Sustained Economic Growth' (2005) 16 JAE 663, 663–87.

[35] Theodore W Schultz, 'Investment in Human Capital' (1961) 51 AER 1, 8–9; Chindo Sulaiman and others, 'Human Capital, Technology, and Economic Growth: Evidence From Nigeria' (2015) 5 SAGE Open 1 <https://journals.sagepub.com/doi/full/10.1177/2158244015615166 >.

[36] Nkem Kumbah, 'STEM Education and African Development' Africa Policy Review, Education and Youth Development; Nigeria Federal Government, *National Policy on Education* (4th edn, Nigeria Federal Government 2004) 35.

[37] Nigeria Federal Government (n 36) 39; African Capacity Building Foundation (ACBF), 'Africa Capacity Report 2017: Building Capacity in Science, Technology and Innovation for Africa's Transformation' (*Africa Portal*, 1 February 2016) 59 <www.africaportal.org/publications/africa-capac

The challenges that the continent is facing in improving the uptake of science, technology, and innovation are also considered. To overcome this challenge and enable Africa to take advantage of the outputs of science and technology, it is proposed that a 'conceptual ecocultural paradigm' be adopted.[38] In terms of the tenets of this paradigm, the design of technology policies must be predicated on the socio-cultural environment and context in which the learner lives and operates.[39] This will ensure that Africa's environment; indigenous scientific and technological principles, theories, and concepts; and values of typically African human feelings gain prominence in shaping technology, hence improving the uptake of science and technology on the continent.[40] If this paradigm is embraced, it will enable the development of technology policies that are based on local needs while simultaneously taking into account pertinent cultural, economic, and political factors and will focus on finding solutions to local needs and priorities. It is thus submitted that future technology policies on the continent incorporate the 'conceptual ecocultural paradigm' principles in policy development.

Finally, the mechanism to promote indigenization of technology acquired in foreign states through utility models is analysed.[41] It is apparent that whereas this form of IP rights could provide a more favourable, accessible, and affordable mechanism for the development of a culture of IP and technological learning in the LDCs, paradoxically it is not yet popular on the continent.[42] This situation ought to be addressed through a very focused strategy and a model strategy that states may adopt to promote the use of utility models is duly provided. The proposed strategy emphasizes the importance of the uptake of utility models as the most appropriate IP category that can fast-track technological learning and the development of absorptive capacities, adaptation, and promotion of local innovation. Accordingly, the proposed strategy highlights some of the mechanisms that can be adopted in terms of institutional frameworks, administration of the rights, awareness

---

ity-report-2017-building-capacity-in-science-technology-and-innovation-for-africas-transformation/> accessed 2 February 2019.

[38] Olugbemiro J Jegede, 'School Science and the Development of Scientific Culture: A Review of Contemporary Science Education in Africa' (1997) 19 IntJSciEduc 1, 16.
[39] ibid.
[40] ibid.
[41] Larry E Westphal, 'Technology Strategies for Economic Development in a Fast Changing Global Economy' (2002) 11 EconInnovNewTechnol 275, 281–82; Nagesh Kumar, 'Intellectual Property Rights, Technology and Economic Development: Experiences of Asian Countries' (2003) 38 EPW 209, 214–15; KS Kardam, 'Utility Model: A Tool for Economic and Technological Development: A Case Study of Japan' (Tokyo Institute of Technology 2007) 52–56 <www.ipindia.gov.in/writereaddata/images/pdf/FinalReport_April2007.pdf> accessed 23 January 2019.
[42] Uma Suthersanen, Graham Dutfield, and Kit Boey Chow (eds), *Innovation Without Patents: Harnessing the Creative Spirit in a Diverse World* (Edward Elgar 2007) 20–21; Nishantha Sampath Punchi Hewage, *Promoting a Second-Tier Protection Regime for Innovation of Small and Medium-Sized Enterprises in South Asia: The Case of Sri Lanka* (1st edn, Nomos Verlagsgesellschaft mbH 2015).

creation, funding, and other government support. The conclusion drawn is that Africa needs to shift its approach with regard to the most appropriate IP right that can steer technological progress, and this study has identified utility models as the right means to achieve that objective.

The latest developments related to the Fourth Industrial Revolution (4IR) were not ignored. The 4IR technologies may be a game-changer in Africa. Therefore, special attention is also dedicated to the need to leverage the opportunities offered by these technologies to enable the continent to catch up on technological progress. Africa appears to be placed in a privileged position to that end in view of the demographic dividend that favours the continent, the relatively lower requirements in terms of infrastructure, and its innovative and creativity capabilities.

C. Conclusion

Africa possesses the ability to reclaim its space in global innovation if the correct interventions are effected at the philosophical, policy, legal, and institutional levels. This book has shown that the innovation and creativity that characterize humankind were not only born in Africa but have continued to be manifest around the world. However, that global phenomenon gained momentum at a rapid pace in those regions that adopted the IP system. If the culprit in the exclusion of Africa from technological progress was the IP system, then this study asserts that, correspondingly, it is within that system that solutions can be found. For Africa to recover the time that has been wasted and embrace technological progress, it needs to take decisive steps to adopt and strategically exploit the IP system. The lived reality is disappointing because, notwithstanding the rhetoric that Africa's future development hinges on knowledge-based and innovation-led economies, the policies that are adopted to drive such development do not provide guidance on the important role that IP can play.[43] This book thus engages extensively with the fundamental contribution that technology can make to the development of states. The book also shows that African states do not have the ability to develop the much-needed technology on their own and are forced to obtain it through the process of technology transfer. Following the rights-based approach, a claim for technology transfer can only be sustained if it is based on rights crystallized in a legal instrument. Arguments contained in this book show that in the current international legal system, the TRIPS Agreement offers the strongest legal basis for that claim through Article 66.2, which requires developed states to provide incentives in order to ignite the dynamics of technology transfer for the benefit of LDCs.

---

[43] African Union, 'Science, Technology and Innovation Strategy for Africa (STISA-2024)' 22–23 <https://au.int/en/documents/20200625/science-technology-and-innovation-strategy-africa-2024> accessed 7 October 2017.

It is in the implementation of Article 66.2 that insurmountable obstacles are found which have prevented the effective transfer of technology to LDCs. The book therefore offers two alternatives: either improving the implementation of Article 66.2 to render the process of technology transfer effective; or adopting an all-encompassing new international legal instrument on technology transfer. The two proposals are not mutually exclusive and contain several recommendations that, if implemented, may transform the face of Africa with regard to innovation. Africa will recover its former glory in innovation. It is time for what went around with 'out of Africa' to come back!

References

'2050 Long-Term Vision for Nigeria (LTV-2050)' <https://unfccc.int/documents/386681> accessed 27 April 2023

'About Us—CARI: Competitive African Rice Initiative' <www.cari-project.org/> accessed 29 April 2021

Adams & Adams, *Adams & Adams Practical Guide to Intellectual Property in Africa* (PULP 2012)

Adams HH, 'African Observers of the Universe: The Sirius Question' in Ivan Van Sertima (ed), *Blacks in Science: Ancient and Modern* (Transaction Publishers 1980)

Adawo M, 'Has Education (Human Capital) Contributed to the Economic Growth of Nigeria?' (2011) 3 JEIF 46

Adedigba A, 'FACT CHECK: Did UNESCO Ever Recommend 26 per Cent Budgetary Allocation to Education?' (*Premium Times Nigeria*, 9 December 2017) <www.premiumtimesng.com/news/headlines/251927-fact-check-unesco-ever-recommend-26-per-cent-budgetary-allocation-education.html> accessed 6 December 2017

Adhikari R, 'The Fourth Industrial Revolution' in Sarah Aneel, Uzma Haroon, and Imrana Niaza (eds), *Corridors of Knowledge for Peace and Development* (SDPI & Sang-e-Meel Publications 2020)

Adusei P, 'Trajectories of Patent-Related Negotiations Affecting Pharmaceuticals and the Politics of Exclusion in Sub-Saharan Africa' (2010) 24 UGLJ 25–86

'Africa Internet Users, 2022 Population and Facebook Statistics' <www.internetworldstats.com/stats1.htm> accessed 12 March 2019

African Capacity Building Foundation (ACBF), 'Africa Capacity Report 2017: Building Capacity in Science, Technology and Innovation for Africa's Transformation' (*Africa Portal*, 1 February 2016) <www.africaportal.org/publications/africa-capacity-report-2017-building-capacity-in-science-technology-and-innovation-for-africas-transformation/> accessed 2 February 2019

African Union, 'Agenda 2063: The Africa We Want' (*Africa Portal*, September 2015) <https://au.int/en/agenda2063/overview> accessed 4 June 2022

—— 'AIDA—Accelerated Industrial Development for Africa' (*Africa Portal*, February 2008) <https://au.int/en/ti/aida/about> accessed 17 January 2017

—— 'Science, Technology and Innovation Strategy for Africa (STISA-2024)' (*Africa Portal*, June 2014) <https://au.int/en/documents/20200625/science-technology-and-innovation-strategy-africa-2024> accessed 7 October 2017

'Agenda 21: Sustainable Development Knowledge Platform' <https://sustainabledevelopment.un.org/outcomedocuments/agenda21> accessed 19 June 2019

Aguele LI and Agwagah UNV, 'Female Participation in Science, Technology and Mathematics (STM) Education in Nigeria and National Development' (2007) 15 JSS 121

Ahrens J, 'Governance and the Implementation of Technology Policy in Less Developed Countries' (2002) 11 EINT 441

Akintimehin A, 'Effects of Human Capital Investment on Economic Growth in Nigeria' (2018) <www.researchgate.net/publication/325107498_EFFECTS_OF_HUMAN_CAPITAL_INVESTMENT_ON_ECONOMIC_GROWTH_IN_NIGERIA> accessed 6 February 2019

Alani J, 'Effects of Technological Progress and Productivity on Economic Growth in Uganda' (2012) 1 PEFin 14
Alavi A, 'African Countries and the WTO Dispute Settlement Mechanism' (2007) 25 DPR 25
Alpern SB, 'Did They or Didn't They Invent It? Iron in Sub-Saharan Africa' (2005) 32 HistAfr 41
Alvares F, *The Prester John of the Indies* (Published for the Hakluyt Society at the University Press 1961)
Amir M, Mehmood B, and Shahid M, 'Impact of Human Capital on Economic Growth with Emphasis on Intermediary Role of Technology: Time Series Evidence from Pakistan' (2012) 6 AJBM 280
Anderson DH, 'Efforts to Ensure Universal Participation in the United Nations Convention on the Law of the Sea' (1993) 42 ICLQ 654
Andreassen B-A, 'On the Normative Core of the Right to Development' (1997) 24 ForumDevStud 179
Anwar MA and Graham M, 'Digital Labour at Economic Margins: African Workers and the Global Information Economy' (2020) 47 ROAPE 95
Arabi KAM and Abdalla SZS, 'The Impact of Human Capital on Economic Growth: Empirical Evidence from Sudan' (2013) 4 ResWorldEcon 43
Aredo D, 'Developmental Aid and Agricultural Development Policies in Ethiopia 1957–1987' (1992) 17 AfrDev 209
Aregay MW, 'Society and Technology in Ethiopia, 1500–1800' (1984) 17 IJES 127
ARIPO, 'Report of the Administrative Council' (1999) Doc ARIPO/AC/XXIII/19
—— 'Status of Operations of Industrial Property Rights' Doc ARIPO/AC/XLII/3
—— 'ARIPO Annual Report 2018'
—— and WIPO, 'Guidelines on Elaboration of Intellectual Property Policy and Strategy for an Effective Use of the IP System by Universities and Research and Development Institutions in Africa' (2018). <https://www.wipo.int/technology-transfer/en/database-ip-policies-universities-details.jsp?id=12194> accessed 30 October 2023
Asante M and Asante K, 'Great Zimbabwe: An Ancient African City-State' (1983) 5 JAC 84
Austen RA and Headrick D, 'The Role of Technology in the African Past' (1983) 26 ASR 163
Ayodele O, 'The New Information Feudalism: Africa's Relationship with the Global Information Society' (2020) 27 SAJIA 67
Azmi I, Maniatis S, and Sodipo B, 'Distinctive Signs and Early Markets: Europe, Africa and Islam' in Alison Firth (ed), *The Prehistory and Development of Intellectual Property Systems* (Sweet & Maxwell 1997)
Badamasi Usman Babangida 'Human Capital Development and Economic Growth in Nigeria' (2022) 2(2) Polac Management Review (PMR) 275
'Bali Action Plan' (adopted 14 March 2008) Doc FCCC/CP/2007/6/Add.1
Banerjee R and Roy SS, 'Human Capital, Technological Progress and Trade: What Explains India's Long Run Growth?' (2014) 30 JAE 15
Bard I, *History of the World: Africa and the Origins of Humans* (Steck-Vaughn 1992)
Barton J, 'Preserving the Global Scientific and Technological Commons' (2003) paper presented to Science and Technology Diplomacy Initiative and the ICTSD- UNCTAD Project on IPRs and Sustainable Development Policy Dialogue on a Proposal for an International Science and Technology Treaty, 11 April 2003
—— and Maskus K, 'Economic Perspectives on a Multilateral Agreement on Open Access to Basic Science and Technology' (2004) 1 SCRIPT-ed 369–87

Bashir A, 'An Analysis of the Existing Mechanism of Transfer of Technology from the Developed to the Developing Countries under the TRIPS Agreement' (Queen Mary University of London 2011) <www.academia.edu/856420/International_Transfer_of_Technology_under_the_TRIPS_Agreement?auto=download> accessed 9 September 2019

Becker L, 'Deserving to Own Intellectual Property' (1993) 68 Chic-KentLR 609

Beitz CR, 'Human Rights as a Common Concern' (2001) 95 APSR 269

Benhabib S, 'The Legitimacy of Human Rights' (2008) 137 Daedalus 94

Berg BL and Lune H, *Qualitative Research Methods for the Social Sciences* (8th edn, Pearson 2011)

Blakeney M, *Legal Aspects of the Transfer of Technology to Developing Countries* (ESC Pub 1989)

Bloom DE and others, 'Higher Education and Economic Growth in Africa' (2014) 1 IJAHE 22

Boon GC, 'Creating a Research and Development Culture in Southeast Asia: Lessons from Singapore's Experience' (1998) 26 AJSS 49

Boserup E, 'Population and Technology in Preindustrial Europe' (1987) 13 PopulDevRev 691

Bossche PV den, 'The Doha Development Round Negotiations on the Dispute Settlement Understanding' (WTO 2003)

Bouët A and Metivier J, 'Is the WTO Dispute Settlement Procedure Fair to Developing Countries?' (International Food Policy Research Institute) (2017) <www.ifpri.org/publication/wto-dispute-settlement-procedure-fair-developing-countries> accessed 29 April 2023

Braga C, 'Development Goes Digital' (2000) 1 GJIA 23

Bray F, 'Technics and Civilization in Late Imperial China: An Essay in the Cultural History of Technology' (1998) 13 Osiris 11

Breitwieser A and Foster N, 'Intellectual Property Rights, Innovation and Technology Transfer: A Survey' (The Vienna Institute for International Economic Studies 2012) 88 <https://wiiw.ac.at/p-2646.html> accessed 24 April 2023

Brookes EH, *Apartheid: A Documentary Study of Modern South Africa* (Routledge 2022)

Budde-Sung A, 'The Invisible Meets the Intangible: Culture's Impact on Intellectual Property Protection' (2013) 117 JBusEthics 345

Burke VL, 'The Rise of Europe' (1994) 20 HJSR 1

Bütler M and Hauser H, 'The WTO Dispute Settlement System: A First Assessment from an Economic Perspective' (2000) 16 J LawEconOrgan 503

Cameron JD, 'Revisiting the Ethical Foundations of Aid and Development Policy from a Cosmopolitan Perspective' in Stephen Brown, Molly den Heyer, and David R Black (eds), *Rethinking Canadian Aid* (2nd edn, University of Ottawa Press 2016)

Carbonell E and others, 'Eurasian Gates: The Earliest Human Dispersals' (2008) 64 JAR 195

Cataldo P, 'Building an Industrial/Academic Alliance: The Two Directions of Technology Transfer' (1989) 29 ET 33

CENTRE TTL, 'Tralac—Trade Law Centre' (*tralac*) <www.tralac.org/> accessed 9 May 2023

Chahale S, 'Cultural Barriers to the Introduction of Western Model Collective Management Systems in Kenya' in Anand Nair, Claudio Tamburrino, and Angelica Tavella (eds), *Master of Laws in Intellectual Property—Collection of Research Papers* (Edizione Scientifiche Italiane 2013)

Chakroun N, 'Using Technology Transfer Offices to Foster Technological Development: A Proposal Based on a Combination of Articles 66.2 and 67 of the TRIPS Agreement' (2017) 20 JWIP 103

Chee WT, 'Workforce Productivity Enhancement and Technological Upgrading in Singapore' (1997) 14 ASEAN EconBul 46

Chete LN and others, 'Industrial Policy in Nigeria: Opportunities and Challenges in a Resource-Rich Country' in Carol Newman and others (eds), *Manufacturing Transformation: Comparative Studies of Industrial Development in Africa and Emerging Asia* (Oxford University Press 2016)

Chivunga M and Tempest A, 'Digital Disruption in Africa: Mapping Innovations for the AfCFTA in Post-COVID Times' (South African Institute of International Affairs 2021) <www.jstor.org/stable/resrep28288> accessed 2 February 2021

Choy CL, 'Singapore's Development: Harnessing the Multinationals' (1986) 8 CSEA 56

Chu D and Skinner EP, *A Glorious Age in Africa: The Story of 3 Great African Empires* (Reprint edn, Africa World Press 1998)

Ciuriak D and Ptashkina M, 'Leveraging the Digital Transformation for Development: A Global South Strategy for the Data-Driven Economy' (*Centre for International Governance Innovation*, 3 April 2019) </www.cigionline.org/publications/leveraging-digital-transformation-development-global-south-strategy-data-driven/> accessed 30 April 2023

Clark N and Juma C, 'Technological Catch-up: Opportunities and Challenges for Developing Countries' <www.academia.edu/58329996/Technological_Catch_Up_Opportunities_and_Challenges_for_Developing_Countries> accessed 24 April 2023

Cohen W and Levinthal D, 'Fortune Favors the Prepared Firm' (1994) 40 ManageSci 227

Cohen WM and Levinthal DA, 'Absorptive Capacity: A New Perspective on Learning and Innovation' (1990) 35 ASQ 128

'Copenhagen Accord' (adopted 30 March 2010) Doc FCCC/CP/2009/11/Add.1

Coppens Y, 'Hominid Evolution and the Emergence of the Genus Homo' (2013) 9 Scripta Varia 1–15

Cornwall A and Nyamu-Musembi C, 'Putting the "Rights-based Approach" to Development into Perspective' (2004) 25 Am Univ Law Rev 1415

Corredoira RA, Goldfarb BD, and Shi Y, 'Federal Funding and the Rate and Direction of Inventive Activity' (2018) 47 ResPol 1777

Cullet P, 'Human Rights and Intellectual Property Protection in the TRIPS Era' (2007) 29 HRQ 403

Cwik B, 'Labor as the Basis for Intellectual Property Rights' (2014) 17 ETMP 681

Daka T, Iyatse G, and Adekoya F, 'Nigeria Abandons Vision 20:2020, Dreams Agenda 2050' *The Guardian Nigeria News—Nigeria and World News* (10 September 2020) <https://guardian.ng/news/nigeria-abandons-vision-202020-dreams-agenda-2050/> accessed 25 April 2023

Daniels L, 'South Africa Approves New IP Policy, with Guidance from UN Agencies' *Intellectual Property Watch* (Geneva, 6 June 2017)

Daspit JJ and D'Souza DE, 'Understanding the Multi-Dimensional Nature of Absorptive Capacity' (2013) 25 JManagIssues 299

'Decision of the Executive Council of the African Union at the Eighth Ordinary Session' (adopted 21 January 2006) Doc EX.CL/Dec.254 (VIII))

'Decision of the TRIPs Council for Extension of the Transition Period under Article 66.1 for Least-Developed Country Members' (adopted 29 November 2005) Doc IP/C/40

'Decision on Strategy for Science Technology and Innovation in Africa 2024' Doc EX.CL/839(XXV) <https://archives.au.int/bitstream/handle/123456789/161/Assembly%20AU%20Dec%20520%20%28XXIII%29%20_E.pdf?sequence=1&isAllowed=y> accessed 12 June 2019

'Declaration on the TRIPS Agreement and Public Health (adopted 2001) Doc WT/MIN(01)/DEC/2' <www.wto.org/english/thewto_e/minist_e/min01_e/mindecl_trips_e.htm> accessed 20 November 2019

Deere-Birkbeck C and Marchant R, 'The Technical Assistance Principles of the WIPO Development Agenda and Their Practical Implementation' (2010) ICTSD Issue Paper No 28

Demont M and others, 'From WARDA to Africa Rice: An Overview of Rice Research for Development Activities Conducted in Partnership in Africa' in Eric Tollens and others (eds), *Realizing Africa's Rice Promise* (CABI Publishing 2013)

Department of Basic Education, South Africa, 'Action Plan to 2019 Towards the Realisation of Schooling 2030' <www.education.gov.za/Portals/0/Documents/Publications/Action%20Plan%202019.pdf?ver=2015-11-11-162424-417> accessed 5 May 2023

—— 'National Strategy for Mathematics, Science and Technology Education in General and Further Education and Training' <www.voced.edu.au/content/ngv%3A3777> accessed 2 June 2019

Derpmann S, 'Solidarity and Cosmopolitanism' (2009) 12 EthicalTheoryMoralPract 303

'Development Agenda for WIPO' <www.wipo.int/ip-development/en/agenda/index.html> accessed 8 July 2018

Directorate-General for Research and Innovation (European Commission) and others, *Social Innovation as a Trigger for Transformations: The Role of Research* (Publications Office of the European Union 2017) <https://data.europa.eu/doi/10.2777/68949> accessed 24 April 2023

Du Preez M-L, '4IR and Water-Smart Agriculture in Southern Africa: A Watch List of Key Technological Advances' (South African Institute of International Affairs 2020) <www.jstor.org/stable/resrep29534> accessed 5 September 2020

Duller H, 'Role of Technology in the Emergence of Newly Industrializing Countries' (1992) 9 ASEAN EcoBul 45

Ekperiware M, Oladeji S, and Yinusa O, 'Dynamics of Human Capital Formation and Economic Growth in Nigeria' (2017) 10 JGRESS 139

Ekpo U, 'Nigeria Industrial Policies and Industrial Sector Performance: Analytical Exploration' (2014) 3 IOSR JEF 1

Elahi S, *Scenarios for the Future: How Might IP Regimes Evolve by 2025? What Global Legitimacy Might Such Regimes Have?* (European Patent Office 2007)

Elumah L and Shobayo P, 'Effect of Expenditures on Education, Human Capital Development and Economic Growth in Nigeria' (2017) 3 NileJBE 40

European Union, 'Report on the Implementation of Article 66.2 of the TRIPS Agreement' (WTO) Doc IP/C/R/TTI/EU/3 <https://docs.wto.org/dol2fe/Pages/FE_Search/FE_S_S009-DP.aspx?language=E&CatalogueIdList=288475,288461,288378,288403,288245,288236,288039,288037,288041,279272&CurrentCatalogueIdIndex=0&FullTextHash=&HasEnglishRecord=True&HasFrenchRecord=True&HasSpanishRecord=True> accessed 30 October 2023

Fage JD and Tordoff W, *A History of Africa* (4th edn, Routledge 2001)

Fanzo J, 'The Nutrition Challenge in Sub-Saharan Africa | United Nations Development Programme' (2012) UNDP Africa Policy Notes 2012-012 <www.undp.org/africa/publications/nutrition-challenge-sub-saharan-africa> accessed 2 January 2019

FAO, 'National Strategy and Action Plan for the Implementation of the Great Green Wall Initiative in Ethiopia' <www.fao.org/faolex/results/details/en/c/LEX-FAOC169534/> accessed 12 January 2018

Fedderke JW and Luiz JM, 'Does Human Capital Generate Social and Institutional Capital? Exploring Evidence from South African Time Series Data' (2008) 60 OEP 649

Federico PJ, 'Origin and Early History of Patents' (1929) 11 JPatOffSoc'y 292

Fernandez V, 'The Finance of Innovation in Latin America' (2017) 53 IRFA 37

Fischer ME, 'Is the WTO Appellate Body a "Constitutional Court"? The Interaction of the WTO Dispute Settlement System with Regional and National Actors' (2009) 40 GJIL 291

Foray D, 'Technology Transfer in the TRIPS Age: The Need for New Types of Partnerships between the Least Developed and Most Advanced Economies' [2009] IPSDS

Forere M, 'Revisiting African States Participation in the WTO Dispute Settlement through Intra-Africa RTA Dispute Settlement' (2013) 6 LDR 155

Forere MA, *The Relationship of WTO Law and Regional Trade Agreements in Dispute Settlement: From Fragmentation to Coherence* (Wolters Kluwer: Kluwer Law International 2015)

Forero-Pineda C, 'The Impact of Stronger Intellectual Property Rights on Science and Technology in Developing Countries' (2006) 35 ResearchPol 808

Fu X, Zanello G, and Owusu G, 'Innovation in Low Income Countries: A Survey Report' (*GOV.UK*, 2014) <www.gov.uk/research-for-development-outputs/innovation-in-low-income-countries-a-survey-report> accessed 12 June 2019

Fukuda-Parr S, 'The Right to Development: Reframing a New Discourse for the Twenty-First Century' (2012) 79 SocRes 839

Galvez E, 'Agro-Based Clusters in Developing Countries: Staying Competitive in a Globalized Economy' (2010) <www.fao.org/sustainable-food-value-chains/library/details/ru/c/267092/> accessed 18 November 2018

Gandenberger C and others, 'Factors Driving International Technology Transfer: Empirical Insights from a CDM Project Survey' (2015) 16(8) ClimPol 1065

Gao X, 'Paleolithic Cultures in China: Uniqueness and Divergence' (2013) 54 CurrAnthrop 358

Garnier J, *Traité d'Economie Politique* (Hard Press Publishing 2019)

Gillwald A, 'Beyond Access: Addressing Digital Inequality in Africa' (*Centre for International Governance Innovation*, 10 March 2017) <www.cigionline.org/publications/beyond-access-addressing-digital-inequality-africa/> accessed 12 March 2019

Glass AJ and Saggi K, 'International Technology Transfer and the Technology Gap' (1998) 55 JDE 369

Godwill EA, 'Science and Technology in Africa: The Key Elements and Measures for Sustainable Development' (2014) 14 GJSFR: G BIO-TECH & GEN 16

Goodall JVL, *My Friends: The Wild Chimpanzees* (1st edn, National Geographic Society 1967)

Goodwin AJH, 'Metal Age or Iron Age?' (1952) 7 SAAB 80

Gordon WJ, 'A Property Right in Self-Expression: Equality and Individualism in the Natural Law of Intellectual Property' (1993) 102 YaleLJ 1533

Gottschalk P, 'Technology Transfer and Benefit Sharing under the Biodiversity Convention' in Hans Henrik Lidgard and others (eds), *Sustainable Technology Transfer: A Guide to Global Aid & Trade Development* (Kluwer Law International 2012)

Götz M, 'Attracting Foreign Direct Investment in the Era of Digitally Reshaped International Production. The Primer on the Role of the Investment Policy and Clusters—The Case of Poland' (2020) 26 JEast-WestBus 131

'Government Expenditure on Education, Total (% of GDP)' (*World Bank Open Data*) <https://data.worldbank.org/indicator/SE.XPD.TOTL.GD.ZS?view=map&year=2017&year_high_desc=false> accessed 7 February 2019

Government of Ghana, 'Seven-Year Development Plan: 1963/64–1969/70' <https://ndpc.gov.gh/media/Ghana_7_Year_Development_Plan_1963-4_1969-70_1964.pdf> accessed 9 November 2018

Government of Kenya, 'Sector Plan for Science, Technology and Innovation 2013–2017' <www.planning.go.ke/wp-content/uploads/2020/11/SECTOR-PLAN-FOR-SCIENCE-TECHNOLOGY-AND-INNOVATION-2013-2017-1.pdf> accessed 6 February 2019

Government of Mozambique, 'Program for the Reduction of Absolute Poverty—PARPA 2001–2005' <https://www.mef.gov.mz/index.php/todas-publicacoes/instrumentos-de-gestao-economica-e-social/estrategia-para-reducao-da-pobreza/parpa-i-2001-2005/7-parpa-i/file?force-download=1> accessed 10 November 2018

—— 'Program for the Reduction of Absolute Poverty 2006–2009' <https://www.portaldogoverno.gov.mz/por/content/download/1765/14370/version/1/file/PARPA_II_aprovado.pdf> accessed 10 November 2018

Government of South Africa, 'Growth, Employment and Redistribution Strategy' (GEAR 1996) <https://www.gov.za/sites/default/files/gcis_document/201409/gear0.pdf> accessed on 30 October 2023

Gready P, 'Rights-Based Approaches to Development: What Is the Value-Added?' (2008) 18 DevPrac 735

Groussot X and Tran-Wasescha T-L, 'TRIPS Article 66(2)—Between Hard Law and Soft Law?' in Hans Henrik Lidgard, Jeffrey Atik, and Tu Thanh Nguyen (eds), *Sustainable Technology Transfer: A Guide to Global Aid & Trade Development* (Kluwer Law International 2012)

Hall BH and Lerner J, 'The Financing of R&D and Innovation' in Bronwyn H Hall and Nathan Rosenberg (eds), *Handbook of the Economics of Innovation*, vol 1 (North Holland 2010)

Hallgren R, 'The UN and the Right to Development' (1990) 22/23 PeaceRes 31

Hamm BI, 'A Human Rights Approach to Development' (2001) 23 HRQ 1005

'Harare Protocol on Patents and Industrial Designs' <https://www.newaripo.online/storage/resources-protocols/1675437439_Harare-Protocol-on-Patents-and-Industrial-Designs-2023.pdf> accessed 12 February 2019

Haug DM, 'The International Transfer of Technology—Lessons That East Europe Can Learn from the Failed Third World Experience' (1992) 5 HarvJL&Tech 209

Hervey T and others, *Research Methodologies in EU and International Law* (Hart Publishing 2011)

Hesse C, 'The Rise of Intellectual Property, 700 B.C.–A.D. 2000: An Idea in the Balance' (2002) 131 Daedalus 26

Hettinger EC, 'Justifying Intellectual Property' (1989) 18 PhilosPublicAff 31

Hewage NSP, *Promoting a Second-Tier Protection Regime for Innovation of Small and Medium-Sized Enterprises in South Asia: The Case of Sri Lanka* (1st edn, Nomos Verlagsgesellschaft mbH 2015)

Hsu G, 'The Competitive Edge: Approaches to Funding Scientific Research' (1995) 18 HarvIntRev 68

Hull G, 'Clearing the Rubbish: Locke, the Waste Proviso, and the Moral Justification of Intellectual Property' (2009) 23 PAQ 67

Jegede OJ, 'School Science and the Development of Scientific Culture: A Review of Contemporary Science Education in Africa' (1997) 19 IJSE 1

Jeju AO and Kentaro KM, 'Does the Utility Model Protection Regime Make Sense in Least Developing Countries? A Comparative Study of the Ethiopian and the Ugandan Utility

Model Protection Regimes and Its Effect on Innovation and Economic Growth' (Africa University 2012)

Joubert G, Barrett A, and Tikly L, 'Approaches to Strengthening Secondary STEM & ICT Education in Sub-Saharan Africa' (2018) Working Paper No 10 <www.bristol.ac.uk/media-library/sites/education/documents/Binder1.pdf> accessed 2 June 2019

Juma C, *Innovation and Its Enemies: Why People Resist New Technologies* (Oxford University Press 2016)

Kaba B and Meso P, 'Benefitting from Digital Opportunity: Do Socio-Economically Advantaged and Disadvantaged Groups React in the Same Ways?' (2019) 22 JGITM 257

Kahn M, *Science, Technology, Engineering and Mathematics (STEM) in South Africa* (Australian Council of Learned Academies 2013)

Kameri-Mbote P, 'Intellectual Property Protection in Africa: Assessment of the Status of Law, Research and Policy Analysis on Intellectual Property Rights in Kenya' (International Environmental Law Research Centre (IELRC) 2005) Working Paper No 2 <http://erepository.uonbi.ac.ke/handle/11295/41242> accessed 14 May 2017

Kanayo O, 'The Impact of Human Capital Formation on Economic Growth in Nigeria' (2013) 4 J Econ 121

Kanja GM, *Intellectual Property Law* (The University of Zambia 2006)

Kardam K, 'Utility Model: A Tool for Economic and Technological Development: A Case Study of Japan' (Tokyo Institute of Technology 2007) <www.ipindia.gov.in/writereaddata/images/pdf/FinalReport_April2007.pdf> accessed 23 January 2019

Kaushik A, 'Dispute Settlement System at the World Trade Organisation' (2008) 43 EPW 26

Kazanjian D, 'The Speculative Freedom of Colonial Liberia' (2011) 63 AmQ 863

Keller W, 'Absorptive Capacity: On the Creation and Acquisition of Technology in Development' (1996) 49 JDE 199

Kendie D, 'The Causes of the Failure of the Present Regime in Ethiopia' (2003) 1 IJES 177

Kessie E and Addo K, 'African Countries and the WTO Negotiations on the Dispute Settlement Understanding' (2007) <www.trapca.org/wp-content/uploads/2019/09/TWP0807_African_Countries_And_The_WTO_Negotiations_On_The_Dispute_Settlement_Understanding-1.pdf> accessed 15 February 2019

Kim L, 'Technology Policies and Strategies for Developing Countries: Lessons from the Korean Experience' (1998) 10 TechnoAnalStrateg 311

Kingiri AN and Fu X, 'Understanding the Diffusion and Adoption of Digital Finance Innovation in Emerging Economies: M-Pesa Money Mobile Transfer Service in Kenya' (2020) 10 InnovDev 67

Kizza JM, 'Mobile Money Technology and the Fast Disappearing African Digital Divide' (2013) 5 AJSTID 373

Knight C, 'In Defence of Cosmopolitanism' (2011) 58 Theoria: JSPT 19

Kongolo T, 'Historical Developments of Industrial Property Laws in Africa' (2013) 5 WIPO J 105

Kumar N, 'Intellectual Property Rights, Technology and Economic Development: Experiences of Asian Countries' (2003) 38 EPW 209

Kumbah N, 'STEM Education and African Development' (21 November 2016) <http://africapolicyreview.com/stem-education-and-african-development/> accessed 7 February 2019

Kuosmanen J, 'Repackaging Human Rights: On the Justification and the Function of the Right to Development' (2015) 11 JGE 303

Kuruk P, 'Protecting Folklore Under Modern Intellectual Property Regimes: A Reappraisal of the Tensions Between Individual and Communal Rights in Africa and the United

States' (1999) 48 AmUniLawRev <https://digitalcommons.wcl.american.edu/aulr/vol48/iss4/2> (30 October 2023)

Larson BA and Anderson M, 'Technology Transfer, Licensing Contracts, and Incentives for Further Innovation' (1994) 76 AJAE 547

Laughon SW, 'Administration Problems in Maryland in Liberia: 1836–1851' (1941) 26 JNH 325

Lawson N, 'Man Not Only Toolmaker' (1964) 85 Science News <www.sciencenews.org/archive/man-not-only-toolmaker> accessed 18 June 2018

Leakey LSB, *Progress and Evolution of Man in Africa* (Oxford 1961)

Lee D, 'Why Big Tech Pays Poor Kenyans to Teach Self-Driving Cars' *BBC News* (3 November 2018) <www.bbc.com/news/technology-46055595> accessed 9 October 2022

Lee J-W, 'Human Capital and Productivity for Korea's Sustained Economic Growth' (2005) 16 JAE 663

Lehman JA, 'Intellectual Property Rights and Chinese Tradition Section: Philosophical Foundations' (2006) 69 JBE 1

Lema R and others, 'Renewable Electrification and Local Capability Formation: Linkages and Interactive Learning' (2018) 117 EnergyPol 326

Lemon A and Stevens L, 'Reshaping Education in the New South Africa' (1999) 84 Geography 222

Lepore J, 'Our Own Devices: Does Technology Drive History?' *The New Yorker* (12 May 2008) <www.newyorker.com/magazine/2008/05/12/our-own-devices> accessed 28 June 2018

Lewis DL, *The Race to Fashoda: European Colonialism and African Resistance in the Scramble for Africa* (1st edn, Weidenfeld Nicolson 1987)

Lidgard HH, 'Assessing Reporting Obligations under TRIPS Article 66.2' in Hans Henrik Lidgard and others (eds), *Sustainable Technology Transfer: A Guide to Global Aid & Trade Development* (Kluwer Law International 2012)

Locke J, *Two Treatises of Government* (Peacock 1689)

Long PO, 'Invention, Authorship, "Intellectual Property," and the Origin of Patents: Notes toward a Conceptual History' (1991) 32 TechCult 846

Lowrey Y and Baumol WJ, 'Rapid Invention, Slow Industrialization, and the Absent Innovative Entrepreneur in Medieval China' (2013) 157 PAPS 1

Lumpkin B, 'The Pyramids: Ancient Showcase of African Technology' (1980) 2 JAC 10

Lundvall B-Å and Lema R, 'Growth and Structural Change in Africa: Development Strategies for the Learning Economy' in Padmashree Gehl Sampath and Banji Oyelaran-Oyeyinka (eds), *Sustainable Industrialization in Africa: Toward a New Development Agenda* (Palgrave Macmillan UK 2016)

Lustig BA, 'Natural Law, Property, and Justice: The General Justification of Property in John Locke' (1991) 19 JReligEthics 119

Lynch BM and Robbins LH, 'Namoratunga: The First Archeoastronomical Evidence in Sub-Saharan Africa' in Ivan Sertima (ed), *Blacks in Science: Ancient and Modern* (Transaction Publishers 1980)

Ma ZF, 'The Effectiveness of Kyoto Protocol and the Legal Institution for International Technology Transfer' (2012) 37 JTT 75

Maak T and Pless NM, 'Business Leaders as Citizens of the World. Advancing Humanism on a Global Scale' (2009) 88 JBE 537

Maguire JW-S, 'Progressive IP Reform in the Middle Kingdom: An Overview of the Past, Present, and Future of Chinese Intellectual Property Law' (2012) 46 IntLawyer 893

Maitra B, 'Investment in Human Capital and Economic Growth in Singapore' (2016) 17 GloBusRev 425

Mansell R, 'Digital Opportunities and the Missing Link for Developing Countries' (2001) 17 OREP 282

Manteaw SO, 'Patent Law in Ghana: Proposals for Change' (2010) 24 UGLJ

Maskus K, 'Encouraging International Technology Transfer' [2004] IPRSD 1

Maskus KE, 'The WIPO Development Agenda' in Neil Weinstock Netanel (ed), *The Development Agenda: Global Intellectual Property and Developing Countries* (Oxford University Press 2008)

—— and Fink C (eds), *Intellectual Property and Development: Lessons from Recent Economic Research* (1st edn, World Bank Publications 2005)

Massa I, 'Technological Change in Developing Countries: Trade-Offs between Economic, Social, and Environmental Sustainability' (2015) MERIT Working Papers No 051

Masters WA, 'Paying for Prosperity: How and Why to Invest in Agricultural Research and Development in Africa' (2005) 58 JIA SIPA 35

Matli W and Ngoepe M, 'Capitalizing on Digital Literacy Skills for Capacity Development of People Who Are Not in Education, Employment or Training in South Africa' (2020) 12 AJSTID 129

May C, 'The Hypocrisy of Forgetfulness: The Contemporary Significance of Early Innovations in Intellectual Property' (2007) 14 RIPE 1

—— and Sell SK, *Intellectual Property Rights: A Critical History* (Lynne Rienner Publishers 2006)

Mboya R, 'Utilization of Industrial Designs and Utility Models in Africa: A Case Study of Kenya' (2021) [Rose Adhiambo Mboya 'Utilization of Industrial Designs and Utility Models in Africa: A Case Study of Kenya'pp. 139 – 154 in Antony Taubman, Martha Chikowore, Wolf R. Meier-Ewert, Tana Pistorius, Sheila Mavis SM. Nyatlo and Bassam Peter Khazin (eds) WIPO-WTO Colloquium Papers: 2018 Africa Edition (Research Papers from the 2018 Regional WIPO-WTO Colloquium for IP Teachers and Scholars in Africa) (2021)]

McBrearty S, 'The Origin of Modern Humans' (1990) 25 Man 129

McCormick D, 'Industrialization Through Cluster Upgrading: Theoretical Perspectives' in Banji Oyelaran-Oyeyinka and Dorothy McCormick (eds), *Industrial Clusters and Innovation Systems in Africa: Institutions, Markets, and Policy* (United Nations University Press 2007)

McMahon J and Young MA, 'The WTO'S Use of Relevant Rules of International Law: An Analysis of the Biotech Case' (2007) 56 ICLQ 907

Mendes APF, Bertella MA, and Teixeira RFAP, 'Industrialization in Sub-Saharan Africa and Import Substitution Policy' (2014) 34 BrazilJPolEcon 120

Mendes de Oliveira M, Santos MC, and Kiker BF, 'The Role of Human Capital and Technological Change in Overeducation' (2000) 19 EconERev 199

Merges RP, *Justifying Intellectual Property:* (Harvard University Press 2011)

Michaels AC, 'International Technology Transfer and TRIPS Article 66.2: Can Global Administrative Law Help Least-Developed Countries Get What They Bargained For?' (2009) 41 GJIL

Michaud P and Vencatachellum D, 'Human Capital Externalities in South Africa' (2003) 51 EDCC 603

Ministry of Industrialization, 'National Industrialization Policy Framework for Kenya 2012–2030: Transforming Kenya into a Globally Competitive Regional Industrial Hub' <https://repository.kippra.or.ke/handle/123456789/1037> accessed 18 June 2017

Moon S, 'Does TRIPS Art. 66.2 Encourage Technology Transfer to LDCs?: An Analysis of Country Submissions to the TRIPS Council (1999–2007) |' (UNCTAD 2008) Policy Briefing 9 <www.eldis.org/document/A42637> accessed 17 March 2018

—— 'Meaningful Technology Transfer to LDCs: A Proposal for a Monitoring Mechanism for TRIPS Article 66.2' (2011) 9 <https://issuu.com/ictsd/docs/technology-transfer-to-the-ldcs> accessed 1 May 2023

Morgera E, Tsioumani E, and Buck M, *Unraveling the Nagoya Protocol: A Commentary on the Nagoya Protocol on Access and Benefit-Sharing to the Convention on Biological Diversity* (Brill Nijhoff 2014)

Mosoti V, 'Does Africa Need the WTO Dispute Settlement System?' in Gregory Shaffer, Victor Mosoti, and Asif H Qureshi (eds), *Towards a Development-supportive Dispute Settlement System in the WTO* (ICTSD 2003) <www.eldis.org/document/A12966> accessed 28 April 2023

Mower JH, 'The Republic of Liberia' (1947) 32 J NegroHist 265

Mowery DC and Oxley JE, 'Inward Technology Transfer and Competitiveness: The Role of National Innovation Systems' (1995) 19 CambJEcon 67

Mozambique, 'Intellectual Property Strategy 2008–2018' <www.aripo.org/resources/mozambique-intellectual-property-strategy-2008-2018/> accessed 16 October 2017

—— 'Agenda 2024: Nation's Vision and Strategies' <https://www.mef.gov.mz/index.php/publicacoes/agenda-2025/83-agenda-2025/file2> accessed 10 November 2018

Muchie M, 'Why Pan-African Education Should Be Promoted' (2015) 553 NewAfr 24

Mutsvairo B and Ragnedda M, 'Comprehending the Digital Disparities in Africa' in Bruce Mutsvairo and Massimo Ragnedda (eds), *Mapping Digital Divide in Africa* (Amsterdam University Press 2019)

Naghavi A, 'Strategic Intellectual Property Rights Policy and North-South Technology Transfer' (2007) 143 RevWorldEcon/Weltwirtschaftliches Archiv 55

Naudé W, 'Brilliant Technologies and Brave Entrepreneurs: A New Narrative for African Manufacturing' (2018) IZA Discussion Papers 11941, Institute of Labor Economics (IZA)

Ncube C, 'The Development of Intellectual Property Policies in Africa: Some Key Considerations and a Research Agenda' (2013) 1 J IntellectPropRights 1

Ndemo B, 'Effective Innovation Policies for Development: The Case of Kenya' in *The Global Innovation Index 2015: Effective Innovation Policies for Development* (World Intellectual Property Organization 2015)

—— 'Effective Innovation Policies for Development: The Case of Kenya' in INSEAD and WIPO (eds), *The Global Innovation Index 2015: Effective Innovation Policies for Development* (WIPO 2015)

—— and Weiss T, 'Making Sense of Africa's Emerging Digital Transformation and Its Many Futures' (2017) 3 AJM 328

Ndlovu-Gatsheni SJ, 'Genealogies of Coloniality and Implications for Africa's Development' (2015) 40 AfrDev 13

Ndungu MN, 'Promotion and Protection of Utility Models for Economic Development in Kenya' (LLM thesis, Africa University 2015)

Newsome F, 'Black Contributions to the Early History of Western Medicine' in Ivan Van Sertima (ed), *Blacks in Science: Ancient and Modern* (Transaction Publishers 1980)

Nigeria Federal Government, *National Policy on Education* (4th edn, Nigeria Federal Government 2004)

'Nigeria Vision 20:2020' <www.nigerianstat.gov.ng/pdfuploads/Abridged_Version_of_Nigeria%20Vision%202020.pdf> accessed 10 November 2018

'Nigerian Industrial Revolution Plan (NIRP)' <www.nipc.gov.ng/product/nigerian-industrial-revolution-plan-nirp/> accessed 10 November 2018

'Nigeria's Economy: Services Drive GDP, but Oil Still Dominates Exports' (*Africa Check*, 2018) <http://africacheck.org/fact-checks/reports/nigerias-economy-services-drive-gdp-oil-still-dominates-exports> accessed 12 March 2020

Nizamuddin AM, 'Multinational Corporations and Economic Development: The Lessons of Singapore' (2007) 82 IntSocSciJ 149

Nozick R, *Anarchy, State, and Utopia* (Reprint edn, Basic Books 2013)

O'Brien DJ, 'Sollicitudo Rei Socialis' in Thomas A Shannon and David O'Brien (eds), *Catholic Social Thought: The Documentary Heritage* (5th pr edn, Orbis Books 1992)

Odek JO, 'The Kenya Patent Law: Promoting Local Inventiveness or Protecting Foreign Patentees?' (1994) 38 JAL 79

OECD, *The Well-Being of Nations: The Role of Human and Social Capital* (OECD Publishing 2001)

—— 'Digital Economy Outlook 2017' <www.oecd.org/digital/oecd-digital-economy-outlook-2017-9789264276284-en.htm> accessed 19 July 2020

Oeij PRA and others, 'Understanding Social Innovation as an Innovation Process: Applying the Innovation Journey Model' (2019) 101 J BusRes 243

Oguamanam C, 'Local Knowledge as Trapped Knowledge: Intellectual Property, Culture, Power and Politics' (2008) 11 JWIP 29

Okediji RL, 'Africa and the Global Intellectual Property System: Beyond the Agency Model' (2004) 12 AYIL 207

Olawuyi DS, 'From Technology Transfer to Technology Absorption: Addressing Climate Technology Gaps in Africa' (2018) 36 JENRL 61

Oliver RA and Fagan BM, *Africa in the Iron Age: C.500 BC–1400 AD* (Cambridge University Press 1975)

Olubanwo F, Oguntuase O, and Ighodalo B, 'Strengthening Intellectual Property Rights and Protection in Nigeria—Trademark—Nigeria' (11 March 2019) <www.mondaq.com/nigeria/trademark/788714/strengthening-intellectual-property-rights-and-protection-in-nigeria> accessed 15 October 2019

'Only 3 African States Spend More Than 1% of GDP on Research & Development | ReSAKSS' (*Africa Wide*, 19 June 2011) <www.resakss.org/node/6285> accessed 6 May 2023

Organisation of African Unity, 'Monrovia Declaration of Commitment of the Heads of State and Government, of the Organization of African Unity on Guidelines and Measures for National and Collective Self-Reliance in Social and Economic Development for the Establishment of a New International Economic Order' AHG/ST. 3 (XVI) Rev 1 (1979) <https://archives.au.int/handle/123456789/835> accessed 12 December 2019

Osei Tutu J, 'Prioritising Human Development in African Intellectual Property Law' (2016) 26 WIPO J

Osman-Gani AM, 'Human Capital Development in Singapore: An Analysis of National Policy Perspectives' (2004) 6 ADHR 276

O'Toole T, 'The Historical Context' in April A Gordon and Donald L Gordon (eds), *Understanding Contemporary Africa* (3rd edn, Lynne Rienner Publishers 2001)

Ott H, 'The New Montreal Protocol: A Small Step for the Protection of the Ozone Layer, a Big Step for International Law and Relations' (1991) 24 VRÜ/WCL 188

Overseas Development Institute, *Industrialization in Sub-Saharan Africa* (1986)

Ovitt G, 'The Cultural Context of Western Technology: Early Christian Attitudes toward Manual Labor' (1986) 27 TechnoCult 477

Oyelaran-Oyeyinka B, *How Can Africa Benefit from Globalization?: Global Governance of Technology and Africa's Global Exclusion* (African Technology Policy Studies Network 2004)

Oyewunmi A, *Nigeria Law of Intellectual Property* (Unilag Press & Bookshop 2015)

Pappademos J, 'An Outline of Africa's Role in the History of Physics' in Ivan Van Sertima (ed), *Blacks in Science: Ancient and Modern* (Transaction Publishers 1980)

Patel SJ, 'The Technological Dependence of Developing Countries' (1974) 12 J ModAfrStud 1

Pelinescu E, 'The Impact of Human Capital on Economic Growth' (2015) 22 ProcediaEconFin 184

Penrose ET, *The Economics of the International Patent System* (Johns Hopkins Press 1951)

Peukert A, 'The Colonial Legacy of the International Copyright System' in Ute Röchenthaler and Mamadou Diawara (eds), *Copyright Africa: How Intellectual Property, Media and Markets Transform Immaterial Goods* (Sean Kingston Publishing 2012)

Piva M, 'The Economic Impact of Technology Transfer in Developing Countries' (2004) 112 RISS 433

Pogge T, *World Poverty and Human Rights*, vol 19 (Cambridge University Press 2002) <www.cambridge.org/core/journals/ethics-and-international-affairs/article/abs/world-poverty-and-human-rights/A647319E9BEE481BAABCADD0B982D89D> accessed 1 May 2023

Pollitzer E, 'Creating a Better Future: Four Scenarios for How Digital Technologies Could Change the World' (2018) 72 JIA SIPA 75

'Population Pyramids of the World from 1950 to 2100' (*PopulationPyramid.net*) <www.populationpyramid.net/singapore/1965/> accessed 9 May 2023

'Priority Needs for Technical and Financial Cooperation—Communication from Sierra Leone' Doc IP/C/W/499

'Programa Quinquenal Do Governo 2010–2014' <www.portaldogoverno.gov.mz/por/content/download/1959/15690/version/1/file/Plano+Quinquenal+do+Governo+2010-14.pdf> accessed 15 August 2018

'Programa Quinquenal Do Governo 2015–2019' (BR I Série N° 29 de 14 de Abril de 2015) <www.mef.gov.mz/index.php/todas-publicacoes/instrumentos-de-gestao-economica-e-social/programa-quinquenal-do-governo-pqg/doismilequinze-doismiledezanove/797-balanco-do-pqg-2015-2019/file?force-download=1)> accessed 18 September 2019

'Proposal by Argentina and Brazil for the Establishment of a Development Agenda for WIPO' Doc WO/GA/31/11 <www.wipo.int/edocs/mdocs/govbody/en/wo_ga_31/wo_g a_31_11.pdf> accessed 8 July 2018

'Proposal by the African Group (25 September 2002)' Doc TN/DS/W/1

'Proposal on the Implementation of Article 66.2 of the Trade-Related Aspects of Intellectual Property Rights (TRIPS) Agreement—Communication from Cambodia on Behalf of the LDC Group' (WTO 2018) Doc IP/C/W/640 <https://docs.wto.org/dol2fe/Pages/FE_Search/FE_S_S009-DP.aspx?language=E&CatalogueIdList=243589,243337,243336,243182,243183,243179,243200,241809,240388,239456&CurrentCatalogueIdIndex=6&FullTextHash=&HasEnglishRecord=True&HasFrenchRecord=True&HasSpanishRecord=True> accessed 12 March 2019

Pueyo A and others, 'How to Increase Technology Transfers to Developing Countries: A Synthesis of the Evidence' (2012) 12 ClimPol 320

Rath S, 'Science and Technology: A Perspective for the Poor' (1994) 29 EPW 2916

Ray S, 'Technology Transfer and Technology Policy in a Developing Country' (2012) 46 JDA 371

'Request for an Extension of the Transitional Period under Article 66.1 of the TRIPS Agreement for Least Developed Country Members with Respect to Pharmaceutical Products and for Waivers from the Obligation of Articles 70.8 and 70.9 of the TRIPS Agreement—Communication from Bangladesh on Behalf of the LDC Group' Doc IP/C/W/605

Resnik DB, 'A Pluralistic Account of Intellectual Property' (2003) 46 JBE 319

Rice CF, Yayboke EK, and Runde D, 'Kenya Case Study' in Charles F Rice, Erol K Yayboke, and Daniel Runde (eds), *Innovation-Led Economic Growth: Transforming Tomorrow's Developing Economies through Technology and Innovation* (Center for Strategic & International Studies 2017)

Roberts K, 'Securing Australia's Future STEM: Country Comparisons' (Australian Council of Learned Academies) <https://acola.org/wp-content/uploads/2018/12/Consultant-Report-South-Africa.pdf> accessed 4 February 2019

Rodney W and Davis A, *How Europe Underdeveloped Africa* (Verso 2018)

Roffe P and Tesfachew T, 'Revisiting the Technology Transfer Debate: Lessons for the New WTO Working Group' (2002) 6 Bridges, ICTSD

Romer P, 'Human Capital and Growth: Theory and Evidence' (1990) 32 Carnegie-Rochester CSPP 251

Rutenberg I and Makanga L, 'Utility Model Protection in Kenya: The Case for Substantive Examination' [2016] AJIC 19

Rwanda, 'Priority Needs for Technical and Financial Co-Operation: Communication from Rwanda' Doc IP/C/W/548

—— 'Revised Policy on Intellectual Property in Rwanda' <https://www.newaripo.online/storage/resources-member-state-policies/1674822545_phpHJDmGy.pdf> accessed 29 October 2023

'Rwanda Vision 2020' <https://www.minaloc.gov.rw/index.php?eID=dumpFile&t=f&f=18970&token=696d20533851dba849455319db973abfa785cdf4> accessed 7 November 2023

Saidi T and Douglas TS, 'Towards the Socialization of Science, Technology and Innovation for African Development' (2018) 10 AJSTID 110

Samkange SJT and Samkange TM, *Hunhuism or Ubuntuism: A Zimbabwe Indigenous Political Philosophy* (Graham Pub 1980)

Sampath P and Roffe P, 'Unpacking the International Technology Transfer Debate: Fifty Years and Beyond' (2012) <https://papers.ssrn.com/abstract=2268529> accessed 20 June 2021

Schildkrout E and Keim C, 'Mangbetu Ivories: Innovations between 1910 and 1914' (1990) 5 AfriHum

Schmidt T, 'Absorptive Capacity—One Size Fits All? A Firm-Level Analysis of Absorptive Capacity for Different Kinds of Knowledge' (2010) 31 MDE 1

Schmitz H, 'On the Clustering of Small Firms' (1992) 23 IDS Bulletin 64

Schultz TW, 'Investment in Human Capital' (1961) 51 AER 1

Schwab K, 'The Fourth Industrial Revolution: What It Means and How to Respond' (*World Economic Forum*, 14 January 2016) <www.weforum.org/agenda/2016/01/the-fourth-industrial-revolution-what-it-means-and-how-to-respond/> accessed 8 October 2022

Schwab K, Davis N, and Nadella S, *Shaping the Future of the Fourth Industrial Revolution* (Illustrated edn, Currency 2018)

Sell SK, *Power and Ideas: North-South Politics of Intellectual Property and Antitrust* (State University of New York Press 1997)

Sellars J, 'Stoic Cosmopolitanism and Zeno's "Republic"' (2007) 28 HPT 1

Sengupta A, 'On the Theory and Practice of the Right to Development' (2002) 24 HRQ 837

Senter R, 'The Impact of Government Research and Development Spending and Other Factors on State Economic Development' (1999) 23 PAQ 368

Shea J, 'Refuting a Myth About Human Origins' (2011) 99 AmeSci 128

Shore D, 'Steel-Making in Ancient Africa' in Ivan Van Sertima (ed), *Blacks in Science: Ancient and Modern* (Transaction Publishers 1991)

Shrivastava M and Shrivastava S, 'Political Economy of Higher Education: Comparing South Africa to Trends in the World' (2014) 67 HighEduc 809

Shugurov MV, 'TRIPS Agreement, International Technology Transfer and Least Developed Countries' (2015) 2 JARE 72

Sikoyo GM, Nyukuri E, and Wakhungu JW, 'Intellectual Property Protection in Africa: Status of Laws, Research and Policy Analysis in Ghana, Kenya, Nigeria, South Africa and Uganda' (African Centre for Technology Studies 2016) <www.jstor.org/stable/resrep00103.1> 30 October 2023

Simonite T, 'The Pandemic Brings Some African Tech Workers Luxe Lodging' (*WIRED*, 19 May 2020) <www.wired.com/story/pandemic-brings-african-tech-workers-luxe-lodging/> accessed 12 December 2020

Smith MR and Marx L, *Does Technology Drive History?* (The MIT Press 1994)

South Africa, 'Intellectual Property Policy of South Africa—Phase I (2018)' <www.gov.za/documents/intellectual-property-policy-south-africa-%E2%80%93-phase-i-2018-13-aug-2018-0000> accessed 19 March 2019

Stavridis J, 'Marine Technology Transfer and the Law of the Sea' (1983) 36 NWCR 38

Steel W and Evans J, *Industrialization in Sub-Saharan Africa—Strategies and Performance* (World Bank 1985)

Suchman MC, 'Invention and Ritual: Notes on the Interrelation of Magic and Intellectual Property in Preliterate Societies' (1989) 89 ColumLawRev 1264

Sulaiman C and others, 'Human Capital, Technology, and Economic Growth: Evidence From Nigeria' (2015) 5 SAGE Open <https://journals.sagepub.com/doi/full/10.1177/2158244015615166> accessed 30 October 2023

Suthersanen U, 'Utility Models and Innovation in Developing Countries' (2006) <www.academia.edu/27254157/Utility_Models_and_Innovation_in_Developing_Countries> accessed 9 February 2019

—— Dutfield G, and Chow KB (eds), *Innovation Without Patents: Harnessing the Creative Spirit in a Diverse World* (Edward Elgar 2007)

Swainson N, 'State and Economy in Post-Colonial Kenya, 1963–1978' (1978) 12 CJAS/RCEA 357

Sweet CM and Eterovic Maggio DS, 'Do Stronger Intellectual Property Rights Increase Innovation?' (2015) 66 WorldDev 665

Syam N and Tellez VM, *Innovation and Global Intellectual Property Regulatory Regimes: The Tension Between Protection and Access in Africa* (South Centre 2016)

Taubman A, *A Practical Guide to Working with TRIPS* (Oxford University Press 2011)

Taylor MS, 'TRIPS, Trade, and Technology Transfer' (1993) 26 CJE/Revue canadienne d'économie 625

'The 45 Adopted Recommendations under the WIPO Development Agenda' <www.wipo.int/ip-development/en/agenda/recommendations.html> accessed 17 October 2017

'The SMART Rwanda Master Plan 2015–2020' <www.minict.gov.rw/fileadmin/user_upload/minict_user_upload/Documents/Policies/SMART_RWANDA_MASTERPLAN.pdf> accessed 16 November 2018

Thomas C, 'Transfer of Technology in the Contemporary International Order' (1998) 22 FordhamIntLJ 2096

Thornton J, 'Early Kongo-Portuguese Relations: A New Interpretation' (1981) 8 HistAfr 183

—— and Mosterman A, 'A Re-Interpretation of the Kongo-Portuguese War of 1622 According to New Documentary Evidence' (2010) 51 JAH 235

Tibebu T, 'Ethiopia: The "Anomaly" and "Paradox" of Africa' (1996) 26 JBS 414

Todorova G and Durisin B, 'Absorptive Capacity: Valuing a Reconceptualization' (2007) 32 AMR 774

Tomlinson S and others, *Innovation and Technology Transfer Framework for a Global Climate Deal An E3G Report with Contributions from Chatham House* (Chatham House 2008)

Toynbee A, *A Study of History*, vol 2 (G Cumberlege, Oxford University Press 1946)

'Trade-Related Aspects of Intellectual Property Rights—Welcome to the e-TRIPS Gateway' <https://e-trips.wto.org/> accessed 9 May 2023

Tripp AM, 'Development and the New Rights-Based Approaches in Africa' (2009) 36 ROAPE 279

Tuniz C, Manzi G, and Caramelli D, *The Science of Human Origins* (Routledge 2016)

Udombana NJ, 'The Third World and the Right to Development: Agenda for the Next Millennium' (2000) 22 HRQ 753

Ugal D and Betiang P, 'Challenges for Developing Human Capital in Nigeria: Global-Local Connection' (2009) <www.researchgate.net/publication/228313475_Challenges_for_Developing_Human_Capital_in_Nigeria_Global-Local_Connection> accessed 1 February 2019

Ujeke B, 'The Challenges of Mathematics in Nigeria's Economic Goals of Vision 2010: Implication for Secondary School Mathematics' (1997)

Ulku H, 'Technological Capability, Innovation, and Productivity in Least-Developed and Developing Countries' in Augusto López-Claros (ed), *The Innovation for Development Report 2010–2011: Innovation as a Driver of Productivity and Economic Growth* (Palgrave Macmillan UK 2011)

United Nations, United Nations A/RES/71/251 Resolution adopted by the General Assembly on 23 December 2016 Establishment of the Technology Bank for the Least Developed Countries

UNCTAD, *Handbook on the Acquisition of Technology by Developing Countries* (United Nations 1978)

—— *Compendium of International Arrangements on Transfer of Technology: Relevant Provisions in Selected International Arrangements Pertaining to Transfer of Technology* (United Nations 2001)

—— *Transfer of Technology* (United Nations 2001)

—— *Transfer of Technology and Knowledge-Sharing for Development: Science, Technology and Innovation Issues for Developing Countries* (2014)

—— 'Digital Economy Report 2019—Value Creation and Capture: Implications for Developing Countries' <https://unctad.org/publication/digital-economy-report-2019> accessed 12 January 2020

UNCTAD Secretariat, 'Draft International Code of Conduct on the Transfer of Technology' <https://digitallibrary.un.org/record/86199> accessed 13 May 2018

UNECA, 'Assessing Regional Integration in Africa VII: Innovation, Competitiveness and Regional Integration' <https://repository.uneca.org/handle/10855/23013> accessed 8 October 2017

UNESCO, 'Education 2030: Luncheon Declaration and Framework for Action: Towards Inclusive and Equitable Quality Education and Lifelong Learning for All' <http://uis.une

sco.org/en/document/education-2030-incheon-declaration-towards-inclusive-equitable-quality-education-and> accessed 14 February 2019
—— 'Global Education Monitoring Report, 2019: Migration, Displacement and Education: Building Bridges, Not Walls' (UNESCO 2019) <https://unesdoc.unesco.org/ark:/48223/pf0000265866> accessed 2 January 2020
—— 'How Much Does Your Country Invest in R&D?' <www.uis.unesco.org/_LAYOUTS/UNESCO/research-and-development-spending/index-en.html> accessed 9 February 2019
'UNFCCC Technology Mechanism' <https://unfccc.int/ttclear/support/technology-mechanism.html> accessed 13 June 2019
UNICEF, 'Education Budget—South Africa 2017/2018' <www.unicef.org/esa/media/2536/file/UNICEF-South-Africa-2018-2019-Education-Budget-Brief.pdf> accessed 2 June 2019
Unknown, 'Liberia' (1909) 3 AJIL 958
Uvin P, 'From the Right to Development to the Rights-Based Approach: How "Human Rights" Entered Development' (2007) 17 DevPra 597
van der Heiden P and others, 'Necessitated Absorptive Capacity and Metaroutines in International Technology Transfer: A New Model' (2016) 41 JET-M 65
von Grebmer K and others, 'Global Hunger Index 2018: Forced Migration and Hunger' (2018) <https://developmenteducation.ie/resource/global-hunger-index-2018-forced-migration-and-hunger/> accessed 17 January 2019
Van Sertima IEBV, *Blacks in Science: Ancient and Modern* (Transaction Publishers 1991)
Waibel M and Alford WP, 'Technology Transfer' in R Wolfrum (ed), *Max Planck Encyclopedia of Public International Law* (Oxford University Press 2011)
Watal J and Caminero L, 'Least-Developed Countries, Transfer of Technology and the TRIPS Agreement' (2018) <www.wto.org/english/res_e/reser_e/ersd201801_e.pdf> accessed 24 April 2018
Wekundah JM, 'A Study on Intellectual Property Environment in Eight Countries: Swaziland, Lesotho, Mozambique, Malawi, Tanzania, Uganda, Kenya and Ethiopia' (African Technology Policy Studies Network (ATPS) 2012) African Technology Policy Studies Network Working Paper Series No 66 <www.africaportal.org/publications/a-study-on-intellectual-property-environment-in-eight-countries-swaziland-lesotho-mozambique-malawi-tanzania-uganda-kenya-and-ethiopia/> accessed 12 January 2019
Westphal LE, 'Technology Strategies for Economic Development in a Fast Changing Global Economy' (2002) 11 EconInnovNewTechnol 275
White L, 'Technology and Invention in the Middle Ages' (1940) 15 Speculum 141
Whitney E, 'Paradise Restored. The Mechanical Arts from Antiquity through the Thirteenth Century' (1990) 80 TAPS 1
WIPO, 'Intellectual Property Needs and Expectations of Traditional Knowledge Holders (WIPO Report on Fact-Finding Missions on Intellectual Property and Traditional Knowledge (1998–1999))' (WIPO 2001) <www.wipo.int/publications/en/details.jsp?id=283> accessed 17 May 2018
—— 'Innovation and Technology Transfer Support Structure for National Institutions (Recommendation 10)' (2010) Doc CDIP/3/INF/2/STUDY/VII/INF/1 <https://dacatalogue.wipo.int/projects/DA_10_03> accessed 30 October 2023
—— 'Policy Guide on Alternatives in Patent Search and Examination' <www.wipo.int/edocs/pubdocs/en/wipo_pub_guide_patentsearch.pdf> accessed 18 October 2017
—— 'Integrating Intellectual Property into Innovation Policy Formulation in Rwanda' (2015) <www.wipo.int/publications/en/details.jsp?id=3943> accessed 16 October 2017

—— 'Report on the WIPO Expert Forum on International Technology Transfer' (2015) Doc CDIP/15/5 <www.wipo.int/meetings/en/doc_details.jsp?doc_id=298371> accessed 26 August 2018

—— 'Methodology for the Development of National Intellectual Property Strategies—Tool 1: The Process' <https://www.wipo.int/edocs/pubdocs/en/intproperty/958/wipo_pub_958_1.pdf> accessed 8 April 2019

—— 'World Intellectual Property Indicators 2018' (2018) <www.wipo.int/publications/en/details.jsp?id=4369> accessed 12 February 2019

Wong C and others, 'From Reality to Law: Sustainable Technology Transfer—An Outlook' in Hans Henrik Lidgard, Jeffery Atik, and Dr Tú Thanh Nguyễn (eds), *Sustainable Technology Transfer: A Guide to Global Aid and Trade Development* (Kluwer Law International 2012)

Wopereis M and others, 'Realizing Africa's Rice Promise: Priorities for Action' in M Wopereis and others (eds), *Realizing Africa's Rice Promise* (CABI Publishing 2013)

World Bank, *Global Economic Prospects 2008: Technology Diffusion in the Developing World (Vol 2)* (World Bank 2008) <https://documents.worldbank.org/en/publication/documents-reports/documentdetail> accessed 30 October 2023

—— *Accelerating Catch-up: Tertiary Education for Growth in Sub-Saharan Africa* (World Bank 2009) <http://hdl.handle.net/10986/2589> accessed 15 January 2019

—— *Financing Higher Education in Africa* (World Bank 2010)

'World Population Dashboard—Nigeria' (United Nations Population Fund) <www.unfpa.org/data/world-population/NG> accessed 17 June 2017

'World Population Prospects—Population Division—United Nations' <https://population.un.org/wpp/> accessed 9 May 2023

WTO, 'Ministerial Conferences—Doha 4th Ministerial Conference—Implementation-Related Issues and Concerns' (2001) Doc WT/MIN/(01)/17 <www.wto.org/english/thewto_e/minist_e/min01_e/mindecl_implementation_e.htm> accessed 19 June 2018

—— 'Implementation of Article 66.2 of the TRIPS Agreement—Decision of the Council for TRIPS' (adopted 19 February 2003) Doc IP/C/28

—— 'Proposed Format for Reports Submitted by the Developed Country Members under Article 66.2' (adopted 3 October 2011) Doc IP/C/W/561

—— 'Minutes of the Session of the TRIPS Council (24–25 October and 17 November 2011)' (2012) Doc IP/C/M/67 <https://docs.wto.org/dol2fe/Pages/FE_Search/FE_S_S009-DP.aspx?language=E&CatalogueIdList=42186,38092,99228,99226,99746,108518,103728,96161,98545,96162&CurrentCatalogueIdIndex=0&FullTextSearch=> accessed 12 March 2019

—— 'Minutes of the Session of TRIPs Council (10–11 October 2013)' Doc IP/C/M/74/Add.1

—— 'Special Session of the Dispute Settlement Body—Report by the Chairman (30 January 2015)' Doc TN/DS/26

—— 'Extension of the Transition Period Under Article 66.1 of the TRIPS Agreement for Least Developed Country Members for Certain Obligations with Respect to Pharmaceutical Products (adopted 6 November 2015) Doc IP/C/73 <www.wto.org/english/tratop_e/trips_e/art66_1_e.htm> accessed 16 November 2018

—— 'Minutes of the Session of the TRIPS Council (8–9 November 2016)' Doc IP/C/M/83/Add.1

—— 'Minutes of the Council for TRIPS—Communication by the Delegation of Benin (27–28 February 2018)' Doc IP/C/M/89/Add.1

—— 'Minutes of the Session of the TRIPS Council (27 February 2018)' Doc IP/C/M/88/Add.1

—— 'Intellectual Property (TRIPS)—Notifications: Contact Points' <www.wto.org/english/tratop_e/trips_e/trips_notif5_art69_e.htm> accessed 29 April 2021

—— 'Special and Differential Treatment Provisions' <www.wto.org/english/tratop_e/devel_e/dev_special_differential_provisions_e.htm> accessed 9 August 2019

Wubneh M, 'State Control and Manufacturing Labor Productivity in Ethiopia' (1990) 24 JDA 311

Yakohene AB, 'Overview of Ghana and Regional Integration: Past, Present and Future' in Friedrich Ebert Stiftung (ed), *Ghana: In Search of Regional Integration Agenda* (Friedrich Ebert Stiftung 2009)

Yankey GS-A, *International Patents and Technology Transfer to Less Developed Countries: The Case of Ghana and Nigeria* (Avebury 1987)

Yao Kouame C, Kalimili BBN, and Pirlea F, 'WDI—Many African Economies Are Larger than Previously Estimated' (2019) <https://datatopics.worldbank.org/world-development-indicators/stories/many-economies-in-ssa-larger-than-previously-thought.html> accessed 18 June 2021

Yomoah DA, 'Lions Go Digital: The Internet's Transformative Potential in Africa' (McKinsey and Company 2013) <https://africanarguments.org/2013/12/lions-go-digital-the-internets-transformative-potential-in-africa-we-profile-the-newest-africa-report-from-mckinsey-by-doreen-akiyo-yomoah/> accessed 12 April 2019

Young S and Lan P, 'Technology Transfer to China through Foreign Direct Investment' (1997) 31 RegStud 669

Young T, Cole J, and Denton D, 'Improving Technological Literacy' (2002) 18 IssuesSciTechnol 73

Yu P, 'A Tale of Two Development Agendas' (2009) 35 ONULawRev 465

Zahra SA and George G, 'Absorptive Capacity: A Review, Reconceptualization, and Extension' (2002) 27 AMR 185

Zambia, 'Revised National Intellectual Property Policy 2020' <https://www.pacra.org.zm/wp-content/uploads/2021/08/RevisedNationalIntellectualPropertyPolicy.pdf> accessed 29 October 2023

Zangato É and Holl AFC, 'On the Iron Front: New Evidence from North-Central Africa' (2010) 8 JAfrArchaeol 7

Zaslavsky C, 'The Yoruba Number System' in Ivan Van Sertima (ed), *Blacks in Science: Ancient and Modern* (Transaction Publishers 1991)

Zeng DZ (ed), *Knowledge, Technology, and Cluster-Based Growth in Africa* (World Bank Publications 2008)

Zhou KZ and Wu F, 'Technological Capability, Strategic Flexibility, and Product Innovation' (2010) 31 SMJ 547

Zimbabwe, 'National Intellectual Property Policy and Implementation Strategy 2018–2022' <https://www.newaripo.online/storage/resources-member-state-policies/1674822688_phpRksdBI.pdf> accessed 29 October 2023

WTO Marrakesh Agreement Establishing the World Trade Organization, Annex 1C, 1869 UNTS 299 (opened for signature 15 April 1994, entered into force 1 January 1995), Agreement on Trade-Related Aspects of Intellectual Property Rights (TRIPS)

Convention on Biological Diversity (opened for signature 5 June 1992, entered into force 29 December 1993) 1760 UNTS 79

International Covenant on Economic, Social and Cultural Rights (adopted 6 December 1966, entered into force 3 January 1976) GA. Resolution 2200A (XXI), UNTS 999

Korea Act 5080, 29 December 1995

Kenya Science, Technology and Innovation Act 2013

Kyoto Protocol to the United Nations Framework Convention on Climate Change (opened for signature 11 December 1997, entered into force on 16 February 2005) 2303 UNTS 162

The Agreement Establishing the Advisory Centre on WTO Law, Annex II (signed 13 November 1999, entered into force 15 July 2001) Doc WT/GC/W/446

The Charter of Economic Rights and Duties of States of 1974 (adopted 12 December 1974) GA Res 3281(XXIX), UN GAOR, 29th Sess, Supp. No 31' 50.

The Montreal Protocol on Substances that Deplete the Ozone Layer 16 September 1987 (opened for signature of 1987, entered into force 1 January 1989) 1562 UNTS 408

Understanding on the Rules and Procedures Governing the Settlement of Disputes (opened for signature 15 April 1994, entered into force 1 January 1995) 1869 UNTS 401

United Nations Agreement Relating to the Implementation of Part XI of the United Nations Convention on the Law of the Sea (adopted 10 December 1982, entered into force 16 November 1994) UNTS 1836

United Nations Convention on the Law of the Sea (opened for signature 10 December1982, entered into force 16 November 1994) 1833 UNTS 397

United Nations Declaration on the Right to Development (adopted 1986) A/RES/41/128

United Nations Framework Convention on Climate Change (opened for signature 9 May 1992, entered into force 21 March 1994) 1771 UNTS 107

United Nations Programme of Action on the Establishment of a New International Economic Order (adopted 1 May 1974) Doc A/RES/S-6/3202

United Nations Resolution on the establishment of the New International Economic Order (adopted 1 May 1974) Doc A/RES/S-6/3201

Vienna Declaration and Programme of Action (adopted 25 June 1993) UN Doc A/CONF.157/23

Index

For the benefit of digital users, indexed terms that span two pages (e.g., 52–53) may, on occasion, appear on only one of those pages.

Boxes are indicated by *b* following the page number

absorptive capacity *see* technology absorption and adaptation capabilities
Accelerated Industrial Development of Africa (AIDA) 12, 13
Access to Basic Science and Technology (ABST) proposal for WTO Multilateral Agreement 87–88, 102–4, 149–50
accessing and exploiting technology for development
   importance of technology for development 3–5
   initiatives to promote technological development in Africa 9–13
   legal framework to promote technology transfer at global level 13–16
   technology development in Africa 6–9
acquisition of foreign technologies *see* foreign technologies, acquisition and indigenization of
adaptation capabilities *see* technology absorption and adaptation capabilities
Adelard of Bath 23–24
Advisory Centre for Technology Transfer and Innovation (WTO), proposal 132–33
Advisory Centre on WTO Law (ACWL) 132–33, 137
Advisory Service on Transfer and Development of Technology (UNCTAD) 132–33
Africa Rice Project 120–21, 126
African Regional Intellectual Property Organization (ARIPO)
   effective use of IP system by universities and R&D institutions in Africa 172–73
   Harare Protocol on Patents and Industrial Designs of 1982 185–86
   utility models, use of 185–86, 188–89
African Union
   Consolidated Plan of Action for Science, Technology and Innovation 2005 169
   initiatives for promoting technology development 12

Science, Technology and Innovation Strategy for Africa (STISA-2024) 12, 169, 172–73, 182
Agenda 21 152–53
Agenda 2063 12
Agreement on Trade-Related Issues of Technology Transfer and Innovation (TRITTI), proposal for
   applicable law 150–51
   conclusions and recommendations 160–61, 213–14
   dispute settlement 144–45, 153–55
   generally 205
   introduction 143–45
   legal character 144–45, 152–53
   private sector collaboration 144–45, 158–60, 210
   rationale for establishment of 145–47
   scope 144
   technical assistance 144–45, 155–56
   WTO as host organisation 144–45, 156–58
agriculture
   4IR technologies, exploitation of 199
Angola
   Sona geometry (symmetrical and nonlinear) 24
animal power
   European use of 33
applicable law
   Agreement on Trade-Related Issues of Technology Transfer and Innovation, proposal for 150–51
   Draft International Code of Conduct for the Transfer of Technology 150–51
art, development of
   Ife bronze heads 19–20, 22, 36
   ivory carving 36
   Nok figurines 19–20, 22, 36
   support from ruling elite 36–37
artificial intelligence
   4IR technologies, exploitation of 197–98, 199
artisans, sharing of knowledge between 31–32

Asia
  cultural behaviour
    technological conservatism and 30
  East Asia
    capacity to absorb technology
      transferred 164
    utility models in East Asian states 186–88
  IP protection and promotion
    knowledge, individual ownership of 42–43
    legal framework 41–42
    political environment 41
  world's scientific and technological progress,
      contribution to 30–31
    China 30–31
    compass 30–31
    entrepreneurial class, absence of 31
    farming innovations 30–31
    gunpowder 30–31
    high-level equilibrium trap 31
    printing 30–31
astronomy 24, 25–26, 27
awareness creation and utility models 192b

Bali Action Plan 152–53
Barton, John
  proposed Multilateral Agreement on Access
    to Basic Science and Technology 102–3,
    104
binary logic of mathematical calculation 24
Biological Diversity Convention 1992
  technology transfer, and 95, 100–1, 149–50,
    153–55
biotechnology
  4IR technologies, exploitation of 197–98, 199
blacksmiths' workshops 35–36
Blue Skies scenario 197–98
Botswana
  national IP policies and strategies 77–78
  utility models, use of 188–89

calendar, development of 25
Cameroon
  import-substituting industries 70–71
  industrial development policies 8
  iron age technology 35–36
  manufacturing activities 7, 68–69
'capable state'
  obligations to transfer technology 114
capacity to absorb technology transferred see
    technology absorption and adaptation
    capabilities
carbon steel production 24, 35–36
Central Africa
  geometric patterns 24
  textile industry 22

Central African Republic
  iron age technology 19–20, 35–36
charity, duty of see Labour Theory (John Locke)
Charter of Economic Rights and Duties of States
    1974
  obligations to transfer technology 113–14
China
  acquisition and indigenization of foreign
    technologies through utility models 188
  IP protection and promotion 23–43
    Ming Dynasty 41
    neo-Confucianism 41
    state control of publishing 42
    Taoism 41
    toleration of copying 42
  world's scientific and technological progress,
    contribution to 30–31
Christian theology, influence of 33–34
climate change
  technology transfer, and 88, 95, 99–100,
    149–50, 152–55
clusters of technology transfer
  food insecurity, and 121–22
  sector-specific clusters, advantages of 122
  start-ups, generation of 122
  utility models, and 193b
Committee on Development and Intellectual
    Property 105, 106
compass, invention of 30–31
compulsory public service corporations
  knowledge, ownership of 46
Convention on Biological Diversity 1992 14,
    149–50, 153–55
Convention on the Law of the Sea 1982 14, 95,
    98–99, 152
Copenhagen Accord 152–53
copying
  toleration of in Asia 42
cosmopolitanism 15–16
'court art' 36
craft guilds and corporations 33
CRISPR (Clustered Regularly Interspaced Short
    Palindromic Repeats)
  use of 4IR technologies in Africa 199
cultural behaviour
  technological conservatism and 30
Cushites' calendar 25
customary law enforcement mechanisms
  IP protection and promotion in Africa 38
cybercrime
  4IR technologies, exploitation of 200

databases, use of 123
Declaration on the Right to Development 113,
    210–11

INDEX

Declaration on the TRIPS Agreement and Public Health 2001 136–37
Democratic Republic of the Congo
  education, spending on 176
  ivory carving 36
developing states, role of see technology absorption and adaptation capabilities
Development Agenda 2004 (WIPO, proposed) 87–88, 104–7
digital economy
  leveraging to fast-track Africa's innovative capabilities 190–96
    attracting investments 195
    building skills 195
    government support 196
    national strategy 195
    targeted intervention 195–96
    technology acquisition 196
digital technologies
  Digital Transformation Strategy for Africa 2020–2030 201–2
  exploiting 4IR technologies 198–99
    convergence of technologies 198
    digital accretion 198
    digital divide 198
    digital harmony 198
    digital inequality 198
    leveraging 4IR value chains 198–99
Digital Transformation Strategy for Africa 2020–2030 201–2
Diophantus 23–24
dispute settlement
  Agreement on Trade-Related Issues of Technology Transfer and Innovation, proposal for 144–45, 153–55
  Draft International Code of Conduct for the Transfer of Technology 153
  Dispute Settlement Mechanism (WTO-DSM) 134–36, 152–55, 206
    challenges to exploiting potential of DSM 136–38
  Multilateral Agreement on Access to Basic Science and Technology (proposed) 103
  Understanding on Rules and Procedures Governing the Settlement of Disputes 134, 153–54
Dogon of Mali
  astronomical observation 24, 26
Doha Decision on Implementation-Related Issues and Concerns 2001 136–37
Draft International Code of Conduct for the Transfer of Technology
  Agreement on Trade-Related Issues of Technology Transfer and Innovation, proposal for 143, 160–61

  applicable law 150–51
  dispute settlement 153
  introduction 14
  legal character 152
  private sector collaboration 158–59
  promoting innovation in Africa 87–88, 95, 96–98, 111
  scope 144
  stumbling blocks to adoption of 147
  technical assistance 155–56
  WTO as host organisation 156–57
drones
  agriculture, use in 199
  drought management, use in 199
drought management
  use of 4IR technologies 199
duty-bearers 113–15
  'capable state' 114
  Charter of Economic Rights and Duties of States 1974 113–14
  Declaration on the Right to Development 113
  governments, role of 113–15
  privately owned technology 114

East Africa
  Cushites' calendar 25
East Asia
  capacity to absorb technology transferred 164
  utility models in East Asian states 186–88
  see also Asia
economic and social development
  technological developments and 4–5
  technology transfer, lack of 5
education, investment in 173–77
Egypt
  calendar, invention of 25
  days of the week 25
  Pyramids of Egypt 19, 23
  time of day 25
enabling environment see technology absorption and adaptation capabilities
enforcement mechanisms
  challenges to exploiting potential of DSM 136–38
  WTO Dispute Settlement Mechanism (DSM) 134–36
engineering, enabling exploitation of 178–81
Eswatini
  proto-maths 24
Ethiopia
  exploitation of science, technology, engineering, and mathematics 179, 180–81
  lithic technology 19
  utility models, use of 188–89, 190

Euclid 23–24
Europe
  IP protection and promotion
    knowledge, individual ownership of 46–48
    legal framework 45–46
    political environment 44–45
  world's scientific and technological progress, contribution to 31–34
    animal power, use of 33
    artisans, sharing of knowledge between 31–32
    Christian theology, influence of 33–34
    craft guilds and corporations 33
    favourable cultural context 34
    machines, use of 33
    medieval period 32–33
    population density 31–32
    universities, establishment of 32
European Patent Office
  scenarios for evolving IP regimes by 2025
    Blue Skies 197–98
    Market Rules 197
    Trees of Knowledge 197
    Whose Game 197
Expert Group on Technology Transfer (EGTT) 156–57
extension of westerns-style IP system into Africa
  conclusions 82–83
  introduction 51–52
  IP systems of non-colonized African states 60–67
    context 60
    Ethiopia 63–67
    Liberia 61–63
  post-independence implementation of IP systems in Africa 67–81
  pre-colonial extension 53–60
    administration structures 59–60
    modalities of introduction of the IP systems in Africa 56–59
    protection of knowledge and innovation 53
    rationale of the introduction of western-style IP systems in Africa 54–56

farming innovations 30–31
Fibonacci (Leonardo Pisano) 23–24
financial support for innovation
  IP protection in Europe 45
fiscal incentives 115–16
  see also incentives
food insecurity
  clusters of technology transfer, and 121–22
foreign technologies, acquisition and indigenization of
  conclusions and recommendations 217–18
  National Strategy to Promote the Use of Utility Models 191*b*–93*b*
  utility models as a tool for indigenization of foreign technologies 182–86
  utility models in Africa 188–90
  utility models in East Asian states 186–88
  see also technology absorption and adaptation capabilities
Fourth Industrial Revolution (4IR) technologies 197–202
  agriculture, use of technologies in 199
  challenges facing LDCs
    absorptive capacities 200
    accessibility 200
    affordability 200–1
    application 201
    cybercrime 200
    unemployment caused by automation 200
  conclusions and recommendations 218
  Digital Transformation Strategy for Africa 2020–2030 201–2
  digital technologies, and 198–99
    convergence of technologies 198
    digital accretion 198
    digital divide 198
    digital harmony 198
    digital inequality 198
    leveraging 4IR value chains 198–99
  EPO scenarios for evolving IP regimes
    Blue Skies 197–98
    Market Rules 197
    Trees of Knowledge 197
    Whose Game 197
  genetics (biotechnology) 197–98, 199
  government intervention 201
  intra-African trade 200
  manufacturing 199–200
  nanotechnology, advances in 197–98
  robotics (artificial intelligence) 197–98, 199
fragmentation 14–15
franchises
  IP protection and promotion in Europe
    legal framework 45
    political environment 44

Gambia, The
  post-independence industrialization policies 68
  utility models, use of 188–89
Gao, university in 27
genetic resources, access to 100–1
genetics
  4IR technologies, exploitation of 197–98, 199

Ghana
 Ghana Research Council 168
 iron age technology 19–20
 utility models, use of 188–89
government intervention
 4IR technologies, exploitation of 201
guilds
 development of
  Africa 39–40
  Europe 46–48
 influence of in Europe 44
gunpowder, invention of 30–31

Harbison-Myers technical enrolment index 178
harmonization, lack of 14–15
Heron 23–24
high-level equilibrium trap 31
Hippocrates 26–27
human capital, development of 173–77, 216
human energy, reliance on
 technological conservatism and 28–29

Ife bronze heads 19–20, 22, 36
Imhotep and the development of
  medicine 26–27
impact-assessment studies 126–32
 capacity to undertake studies 127
 guidelines for field impact-assessments
  institutional arrangements to undertake
   assessment 129b
  introduction 128b
  objective of assessment 128b
  outputs, format and content of 129b
  outputs, use of 129b–30b
  periodicity 128b
  target of assessment 128b
 incentives, identification of 126–27
 priority needs, identification of 127–32
 relevance at field level 127
 requirement for 126
 Technology Bank for LDCs 132
importation of new arts from abroad
 IP protection in Europe 45
incentives
 Africa Rice Project 120–21, 126
 clusters of technology transfer 121–22
 companies and institutions in developed
  states 115–19
 fiscal incentives 115–16
 identification of 126–27
 information and communication technologies,
  use of 123–24
 mechanism to assess the impact of 119–21
 monitoring mechanism 116–17
 non-tax-related incentives 116
 private enterprise projects 116
 reporting mechanism 116–19
indigenization of foreign technologies *see*
  foreign technologies, acquisition and
  indigenization of
information and communication technologies
 databases, use of 123
 LDC-specific programmes, lack of 123–24
ingenuity of humankind and its protection
 conclusion 48–49
 first manifestations 21–22
 intellectual property, birth of 34–48
  political and legal framework for IP
   protection and promotion 35–48
  requirements for development of IP
   system 34–35
 introduction 19–20
 iron age technology 19–20, 22
 lithic technology 19, 21–22, 27
 textile industry 22
 world's scientific and technological
  progress 22–34
  contribution of Africa 22–30
  contribution of Asia 30–31
  contribution of Europe 31–34
Innovation and Technology Transfer Support
 Structure for National Institutions 105–6
innovation patents *see* utility models
institutional arrangements to support
  implementation
 national monitoring institutions 124–26
 WTO Advisory Centre for Technology
  Transfer and Innovation,
  proposal 132–33
 see also impact-assessment studies
intellectual property protection 15
 African framework
  knowledge, individual ownership of 39–40
  legal framework 37–39
  political environment 35–37
 Asian framework
  knowledge, individual ownership of 42–43
  legal framework 41–42
  political environment 41
 European framework
  knowledge, individual ownership of 46–48
  legal framework 45–46
  political environment 44–45
 requirements for development of IP system
  individual ownership of knowledge 35
  political environment 35
  Statute of Anne 1709 (Copyright Act) 34–35, 48

intellectual property protection (*cont.*)
  technological development 35
  Venetian Statute 1474 34–35, 48
International Covenant on Economic, Social and Cultural Rights 1966
  technology transfer, and 95–96
International Technology Transfer Agreement (WTO, proposed) 88, 107–9
intra-African trade
  4IR technologies, exploitation of 200
IP-like protocols in Africa 37, 38
ivory carving 36

Japan
  acquisition and indigenization of foreign technologies through utility models 187–88
Jenne, university in 26–27

Kenya
  exploitation of science, technology, engineering, and mathematics 180–81
  initiatives for promoting technology development 10
  Namoratunga 25
  National Council for Science and Technology 168
  utility models, use of 188–89
knowledge, individual ownership of
  African context 39–40
    guilds, development of 39–40
  Asian context 42–43
    Confucian ethics 42–43
    Islamic world 43
    Judeo-Christian tradition 43
  European context 46–48
    collective ownership of knowledge 46
    compulsory public service corporations 46
    guilds, development of 46–48
    proprietary attitudes, development of 47
    symbols to mark origin of goods 47
Kyoto Protocol to UN Framework Convention on Climate Change 1992
  technology transfer, and 88, 99–100, 149–50, 152–55

Labour Theory (John Locke)
  reinterpretation of 89–94
    moral claims and entitlements 90
    rights-based approach 92–94, 144, 160–61
    TRIPS Agreement 89, 91, 94
    waste limitation 90, 91–92
least developed countries, role of *see* technology absorption and adaptation capabilities

legal frameworks
  African context 37–39
    customary law enforcement mechanisms 38
    IP-like protocols 37, 38
    magic 37–38
    mythology of risk 38
    taboos 37–38
  Asian context
    state control of publishing 42
    toleration of copying 42
  European context 45–46
    early patents in medieval Europe 45–46
    financial support for innovation 45
    franchises for immigrant cloth workers 45
    importation of new arts from abroad 45
    manufacture and sale of medicines 45
  ingenuity of humankind and its protection 35–48
  promoting innovation through technology transfer
    Convention on Biological Diversity 1992 95, 100–1, 149–50, 153–55
    fragmentation in 95–101
    generally 13–16
    Kyoto Protocol to UN Framework Convention on Climate Change 1992 88, 99–100, 149–50, 152–55
    Montreal Protocol on Substances that Deplete the Ozone Layer 1987 95, 101
    Nagoya Protocol on Access to Genetic Resources 100–1
    proposed reforms 102–9
    UN Convention on the Law of the Sea 1982 95, 98–99, 152
    UN Framework Convention on Climate Change 1992 88, 95, 99–100, 149–50, 152–55
    UN International Covenant on Economic, Social and Cultural Rights 1966 95–96
    UNCTAD Draft International Code of Conduct for the Transfer of Technology 1976 87–88, 95, 96–98
    Vienna Convention on Substances that Deplete the Ozone Layer 1985 101
    WIPO Development Agenda 2004 (proposed) 87–88, 104–7
    WTO International Technology Transfer Agreement (proposed) 88, 107–9
    WTO Multilateral Agreement on Access to Basic Science and Technology (proposed) 87–88, 102–4, 149–50
  promoting technology transfer at global level 13–16

utility models 191*b*
Liberia
  education, spending on 176
  extension of westerns-style IP system into Africa 61–63
literacy
  technological conservatism and 29
lithic technology 19, 21–22
Locke's Labour Theory *see* Labour Theory (John Locke)
Long-Term Vision for Nigeria 2050 (LTV-2050) 11–12

magic
  IP protection and promotion in Africa 37–38
Mali
  astronomical observation 24, 26
  identification of priority needs 127–30
  iron age technology 19–20
  Timbuktu, university in 27
manufacturing sector
  4IR technologies, exploitation of 199–200
Market Rules scenario 197
Maskus, Keith
  proposed Multilateral Agreement on Access to Basic Science and Technology 102–3, 104, 109–10
mathematics
  development of 23–24, 27
  enabling exploitation of 178–81, 215
medicine
  development of 26–27
  IP protection in Europe 45
Ming Dynasty
  development of IP in Asia 41
monitoring mechanisms 116–17, 124–26
monopolies
  IP protection in Europe 44
Monrovia Declaration 168–69
Montreal Protocol on Substances that Deplete the Ozone Layer 1987 14, 95, 101
Moors of North Africa
  mathematics, development of 23–24
Mozambique
  national institutions to monitor technology transfers 125–26
  utility models, use of 188–89
Multilateral Environmental Agreements
  dispute settlement mechanisms 153–54
mythology of risk
  IP protection and promotion in Africa 38

Nagoya Protocol on Access to Genetic Resources
  technology transfer, and 100–1

Namoratunga, Kenya 25
nanotechnology
  4IR technologies, exploitation of 197–98
National Commission for Science and Technology
  Kenyan initiatives for promoting technology development 10
National Industrialization Policy Framework for Kenya 2012–2030
  Kenyan initiatives for promoting technology development 10
National Museum for Science and Technology
  Kenyan initiatives for promoting technology development 10
National Research Fund, the National Innovation Agency
  Kenyan initiatives for promoting technology development 10
National Strategy to Promote the Use of Utility Models
  administration of utility models 191*b*–92
  awareness creation 192*b*
  clusters of technology transfer and utility models 193*b*
  funding 192*b*
  government support and resource mobilization 193*b*
  grant of utility models 191*b*–92
  institutional framework 191*b*
  introduction 191*b*
  legal framework 191*b*
  mainstreaming utility models into national priorities of development 192*b*
  promotional activities 192*b*
  synergies between innovators and industry 192*b*
  turning innovations into utility models 192*b*
  see also technology absorption and adaptation capabilities
neo-Confucianism
  development of IP in Asia 41
Nigeria
  artistic works
    Ife bronze heads 19–20, 22, 36
    Nok figurines 19–20, 22, 36
  education
    exploitation of science, technology, engineering, and mathematics 178–79, 180–81
    Harbison-Myers technical enrolment index 178
    National Policy on Education 181–82
    spending on 176–77
  initiatives for promoting technology development

Nigeria (*cont.*)
   Long-Term Vision for Nigeria 2050
     (LTV-2050) 11–12
   Nigeria Industrial Revolution Plan
     (NIRP) 11–12
   Nigeria Vision 20:2020 11–12
   iron age technology 19–20, 22
   Lagos Plan of Action 168–69
   National Council for Scientific and Industrial
     Research 168
   Yoruba of Nigeria
     binary logic of mathematical calculation 24
     system of numbers 24
Nok figurines 19–20, 22, 36
non-tax-related incentives 116
Normalized Difference Vegetation Index
   use of 4IR technologies in Africa 199

Organisation of African Unity (OAU)
   Monrovia Declaration 168–69
ozone layer, depletion of
   technology transfer, and 95, 101

petty patents *see* utility models
philosophical obstacles, overcoming 207–11
physics and physical science, development
   of 24–25, 27
political and military primacy
   technological conservatism and 29
political environment for IP protection and
   promotion
   African context 35–37
     art 36–37
     blacksmiths' workshops 35–36
     carbon steel production 35–36
     iron processing facilities 35–36
     iron smelting 35–36
   Asian context
     Ming Dynasty 41
     neo-Confucianism 41
     Taoism 41
   European context 44–45
     franchises 44
     guilds, influence of 44
     medieval developments 44
     monopolies 44
     public interest 44–45
printing
   invention of 30–31
   IP protection and promotion in
     Asia 42
priority needs
   impact-assessment studies 127–32
private enterprise projects 116

private sector collaboration
   Agreement on Trade-Related Issues of
     Technology Transfer and Innovation,
     proposal for 144–45, 158–60
   Draft International Code of Conduct for the
     Transfer of Technology 158–59
privately owned technology
   obligations to transfer technology 114
Programme Clusters
   AU initiatives for promoting technology
     development 13
promoting innovation through technology
   transfer
   conclusions 109–10
   clusters of technology transfer
     food insecurity, and 121–22
     sector-specific clusters, advantages of 122
     start-ups, generation of 122
   importance of 13–14
   introduction 87–89
   legal framework
     Convention on Biological Diversity
       1992 95, 100–1, 149–50, 153–55
     fragmentation in 95–101
     generally 13–16
     Kyoto Protocol to UN Framework
       Convention on Climate Change 1992 88,
       99–100, 149–50, 152–55
     Montreal Protocol on Substances that
       Deplete the Ozone Layer 1987 95, 101
     Nagoya Protocol on Access to Genetic
       Resources 100–1
     proposed reforms 102–9
     UN Convention on the Law of the Sea
       1982 95, 98–99, 152
     UN Framework Convention on Climate
       Change 1992 88, 95, 99–100, 149–50,
       152–55
     UN International Covenant on Economic,
       Social and Cultural Rights 1966 95–96
     UNCTAD Draft International Code of
       Conduct for the Transfer of Technology
       1976 87–88, 95, 96–98
     Vienna Convention on Substances that
       Deplete the Ozone Layer 1985 101
     WIPO Development Agenda 2004
       (proposed) 87–88, 104–7
     WTO International Technology Transfer
       Agreement (proposed) 88, 107–9
     WTO Multilateral Agreement on Access
       to Basic Science and Technology
       (proposed) 87–88, 102–4, 149–50
   Locke's Labour Theory, reinterpretation
     of 89–94

moral claims and entitlements 90
  rights-based approach 92–94
  TRIPS Agreement 89, 91, 94
  waste limitation 90, 91–92
rights-based approach 87–88, 92–94
  TRIPS Agreement 89, 91, 94, 102–3, 108, 110
  see also Agreement on Trade-Related Issues of Technology Transfer and Innovation (TRITTI); TRIPS Agreement Article 66.2
proto-maths 24
Ptolemy, Claudius 23–24
public interest
  IP protection in Europe 44–45
Pythagorus 26–27

reporting mechanisms 116–19
research and development funding 171–73, 215–16
resource mobilization and utility models 193*b*
rights-based approach
  Locke's Labour Theory, reinterpretation of 92–94, 144, 160–61
risk aversion
  technology development in Africa 6
robotics
  4IR technologies, exploitation of 197–98, 199
Rwanda
  digital economy, leveraging to fast-track innovative capabilities 194
  education, spending on 176
  identification of priority needs 127–30, 131
  iron age technology 19–20
  technology transfer reporting 118
  utility models, use of 188–89

science and technology
  enabling exploitation of 178–81, 215, 216–17
  government institutional frameworks 167–71
    Consolidated Plan of Action for Science, Technology and Innovation 2005 169
    Lagos Plan of Action 168–69
    Ministries of Science and Technology, establishment of 168–70
    Monrovia Declaration 168–69
    Science, Technology and Innovation Strategy for Africa (STISA-2024) 12, 169, 172–73, 182
  Kenyan initiatives for promoting technology development
    Science, Technology and Innovation Act No 28 of 2013 10
    Science, Technology, Innovation Policy and Strategy 2009 10
  socialization or contextualization of 181–82

scientific measurement of quantities 25
sexual division of labour
  human energy, reliance on 29
Sierra Leone
  identification of priority needs 127–31
short-term patents *see* utility models
Singapore
  human capital, development of 174–75
Sirius, astronomical observation of 26
Sona geometry (symmetrical and nonlinear) 24
South Africa
  education
    exploitation of science, technology, engineering, and mathematics 178, 179–81
    Harbison-Myers technical enrolment index 178
    spending on 177
  proto-maths 24
  utility models, use of 188–89
South Korea
  acquisition and indigenization of foreign technologies through utility models 186–87
South Sudan
  education, spending on 176
Special and Differential Treatment
  Multilateral Agreement on Access to Basic Science and Technology (proposed) 103
STISA-2024 *see* Science, Technology and Innovation Strategy for Africa (STISA-2024)
Subsidiary Body for Implementation (SBI) 156–57
Subsidiary Body for Scientific and Technological Advice (SBSTA) 156–57
symbols marking origin of goods
  knowledge, ownership of 47

taboos
  IP protection and promotion in Africa 37–38
Taiwan
  acquisition and indigenization of foreign technologies through utility models 186–87
Tanzania
  carbon steel production 24, 35–36
  identification of priority needs 127–30
  iron smelting 35–36
  utility models, use of 188–89
Taoism
  development of IP in Asia 41
technical assistance
  Agreement on Trade-Related Issues of Technology Transfer and Innovation, proposal for 144–45, 155–56
  Draft International Code of Conduct for the Transfer of Technology 155–56

INDEX

technological conservatism 27–30
  cultural behaviour 30
  human energy, reliance on 28–29
  literacy 29
  political and military primacy 29
  sexual division of labour 29
technological determinism 3–4
technology absorption and adaptation capabilities
  4IR technologies, exploitation of 200
  absorptive capacity
    development of absorptive capacities in Africa 165–66
    government interventions to develop absorptive capacity 166–202
  acquisition and indigenization of foreign technologies
    National Strategy to Promote the Use of Utility Models 191b–93b
    utility models as a tool for indigenization of foreign technologies 182–86
    utility models in Africa 188–90
    utility models in East Asian states 186–88
  conclusions and recommendations 202–3, 214–18
  development of absorptive capacities in Africa 165–66
  government interventions to develop absorptive capacity 166–202
  digital economy 190–96
  exploitation of science, technology, engineering, and mathematics 178–81, 215, 216–17
  foreign technologies, acquisition and indigenization of 182–90, 217–18
  fourth Industrial Revolution 197–202, 218
  government institutional frameworks focused on science, technology, and innovation 167–71
  human capital, development of 173–77, 216
  research and development funding 171–73, 215–16
  socialization or contextualization of science and technology 181–82
  introduction 163–65
Technology Bank for Least Developed Countries 132, 206
technology development in Africa 6–9
  colonial suppression 7
  cultural inertia 6
  current initiatives to promote 9–13
    African Union 12–13
    Kenya 10
    Nigeria 11–12
  decolonization 7–8
  early human development 6
  import substitution 7–8
  industrialization 7–9
  knowledge, lack of 9
  low literacy 6
  marginalization 6
  nationalization 8
  policy and planning, lack of 9
  risk aversion 6
Technology Mechanism on climate change 99, 149–50
technology, science and *see* science and technology
textile industry
  first manifestations of technology 22
Thales of Miletos 24–25
Theon 23–24
Trade Law Centre for Southern Africa (TRALAC) 137
transfer of technology *see* technology transfer
Trees of Knowledge scenario 197
TRIPS Agreement Article 66.2
  clusters of technology transfer
    food insecurity, and 121–22
    sector-specific clusters, advantages of 122
    start-ups, generation of 122
  conclusions 138–39
  duty-bearers 113–15
    'capable state' 114
    Charter of Economic Rights and Duties of States 1974 113–14
    Declaration on the Right to Development 113
    governments, role of 113–15
    privately owned technology 114
  enforcement
    challenges to exploiting potential of DSM 136–38
    WTO Dispute Settlement Mechanism (DSM) 134–36
  guidelines for the field impact-assessment
    generally 128b, 211–13
    institutional arrangements to undertake assessment 129b
    objective of assessment 128b
    outputs, format and content of 129b
    outputs, use of 129b–30b
    periodicity 128b
    target of assessment 128b
  impact-assessment studies 126–32
    capacity to undertake studies 127
    incentives, identification of 126–27
    priority needs, identification of 127–32

INDEX 251

relevance at field level 127
requirement for 126
Technology Bank for LDCs 132, 206
incentives, nature of
 Africa Rice Project 120–21, 126
 clusters of technology transfer 121–22
 companies and institutions in developed states 115–19
 fiscal incentives 115–16
 information and communication technologies, use of 123–24
 mechanism to assess the impact of 119–21
 monitoring mechanism 116–17
 non-tax-related incentives 116
 private enterprise projects 116
 reporting mechanism 116–19
information and communication technologies
 databases, use of 123
 LDC-specific programmes, lack of 123–24
institutional arrangements to support implementation
 impact-assessment studies 126–32
 national monitoring institutions 124–26
 WTO Advisory Centre for Technology Transfer and Innovation, proposal 132–33
introduction 111–12
overview of Article 66.2 112–13
recommendations 211–13

Uganda
 education, spending on 176
 identification of priority needs 127–30
 iron age technology 19–20
 utility models, use of 188–89
unemployment caused by automation 200
United Nations
 Convention on the Law of the Sea 1982 14, 95, 98–99, 152
 Framework Convention on Climate Change 1992 14, 88, 95, 99–100, 149–50, 152–55
 International Covenant on Economic, Social and Cultural Rights 1966 95–96
 Kyoto Protocol to UN Framework Convention on Climate Change 1992 88, 99–100, 149–50, 152–55
 Sustainable Development Goals 152–53
United Nations Conference on Trade and Development (UNCTAD)
 Advisory Service on Transfer and Development of Technology 132–33
 Draft International Code of Conduct for the Transfer of Technology

 Agreement on Trade-Related Issues of Technology Transfer and Innovation, proposal for 143, 160–61
 applicable law 150–51
 dispute settlement 153
 introduction 14
 legal character 152
 private sector collaboration 158–59
 promoting innovation in Africa 87–88, 95, 96–98, 111
 scope 144
 stumbling blocks to adoption of 147
 technical assistance 155–56
 WTO as host organisation 156–57
universities, establishment of
 Africa 27
 Europe 32
utility certificates *see* utility models
utility innovation *see* utility models
utility models
 indigenization of foreign technologies 182–86
 meaning 182–83
 National Strategy to Promote the Use of Utility Models
  administration of utility models 191*b*–92
  awareness creation 192*b*
  clusters of technology transfer and utility models 193*b*
  funding 192*b*
  government support and resource mobilization 193*b*
  grant of utility models 191*b*–92
  institutional framework 191*b*
  introduction 191*b*
  legal framework 191*b*
  mainstreaming utility models into national priorities of development 192*b*
  promotional activities 192*b*
  synergies between innovators and industry 192*b*
  turning innovations into utility models 192*b*
 Paris Convention for the Protection of Industrial Property of 1883 182–83
 trivial patents 184
 use in Africa 188–90
 use in East Asian states 186–88
 see also technology absorption and adaptation capabilities

Venetian Statute 1474 34–35, 48
Vienna Convention on Substances that Deplete the Ozone Layer 1985 101
Vienna Declaration and Programme of Action 1993 210–11

waste limitation
  Locke's Labour Theory, reinterpretation of 90, 91–92
West Africa
  geometric patterns 24
Whose Game scenario 197
World Intellectual Property Organization (WIPO)
  Development Agenda 2004 (proposed) 87–88, 104–7
  effective use of IP system by universities and R&D institutions in Africa 172–73
  World Intellectual Property Indicators
  utility models, use of 185–86, 188–89
World Trade Organization (WTO)
  Advisory Centre for Technology Transfer and Innovation, proposal 132–33, 211–13
  Advisory Centre on WTO Law (ACWL) 132–33, 137
  Declaration on the TRIPS Agreement and Public Health 2001 136–37
  Dispute Settlement Mechanism (WTO-DSM) 134–36, 152–55, 206
    challenges to exploiting potential of DSM 136–38
    Multilateral Agreement on Access to Basic Science and Technology (proposed) 103
    Understanding on Rules and Procedures Governing the Settlement of Disputes 134, 153–54
  Doha Decision on Implementation-Related Issues and Concerns 2001 136–37
  host organization
    Agreement on Trade-Related Issues of Technology Transfer and Innovation, proposal for 144–45, 156–58
    Draft International Code of Conduct for the Transfer of Technology 156–57
  International Technology Transfer Agreement (proposed) 88, 107–9
  Multilateral Agreement on Access to Basic Science and Technology (proposed) 87–88, 102–4

Trade Law Centre for Southern Africa (TRALAC) 137
world's scientific and technological progress
  contribution of Africa
    astronomy 24, 25–26
    calendar, development of 25
    Great Zimbabwe ruins 19, 23
    inputs into scientific and technological progress 22–27
    mathematics, development of 23–24
    medicine 26–27
    physics and physical science, development of 24–25
    Pyramids of Egypt 19, 23
    scientific measurement of quantities 25
    technological 'conservatism' 27–30
    universities 27
  contribution of Asia 30–31
    China 30–31
    compass 30–31
    entrepreneurial class, absence of 31
    farming innovations 30–31
    gunpowder 30–31
    high-level equilibrium trap 31
    printing 30–31
  contribution of Europe 31–34
    animal power, use of 33
    artisans, sharing of knowledge between 31–32
    Christian theology, influence of 33–34
    craft guilds and corporations 33
    favourable cultural context 34
    machines, use of 33
    medieval period 32–33
    population density 31–32
    universities, establishment of 32

Yoruba of Nigeria
  binary logic of mathematical calculation 24
  system of numbers 24

Zimbabwe
  Great Zimbabwe ruins 19, 23
  utility models, use of 188–89